Standards Practice Book

For Home or School
Grade 1

Houghton Mifflin Harcourt

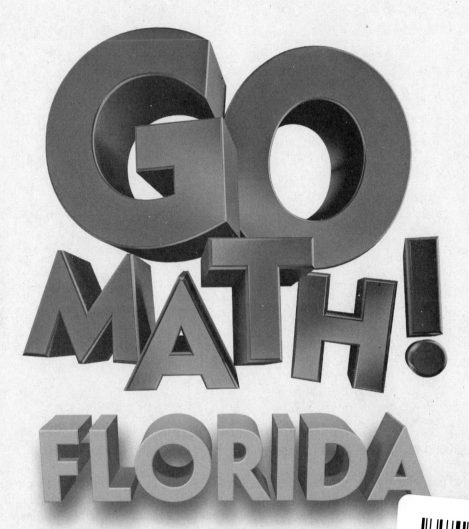

GO MATH! FLORIDA

INCLUDES:

- Home or School Practice
- Lesson Practice and Test Preparation
- English and Spanish School-Home Letters
- Getting Ready for Grade 2 Lessons

Operations and Algebraic Thinking

Developing understanding of addition, subtraction, and strategies
for addition and subtraction within 20

5 Addition and Subtraction Relationships

Domain Operations and Algebraic Thinking

v

Number and Operations in Base Ten

Developing understanding of whole number relationships and place value, including grouping in tens and ones

© Houghton Mifflin Harcourt Publishing Company

8 Two-Digit Addition and Subtraction

Domains Operations and Algebraic Thinking
Number and Operations in Base Ten

Measurement and Data

Developing understanding of linear measurement and measuring lengths as iterating length units

9 Measurement

Domain Measurement and Data

10 Represent Data

Domain Measurement and Data

Geometry

Reasoning about attributes of, and composing and decomposing geometric shapes

End-of-Year Resources

Getting Ready for Grade 2

These lessons review important skills and prepare you for Grade 2.

Table of Contents
Florida Lessons

© Houghton Mifflin Harcourt Publishing Company

School-Home Letter

Dear Family,

My class started Chapter 1 this week. In this chapter, I will learn to add numbers up to ten, to write addition sentences in different ways, and to use pictures to help me add.

Love, _____

Vocabulary

plus (+) part of an addition sentence that means "to add to"

$$\overset{plus}{1 + 3} = 4$$

sum the answer to an addition sentence

$$2 + 4 = ⑥$$

zero a number that means none; if you add zero to any number, the number does not change

$$1 + 0 = 1$$

Home Activity

Use 10 checkers, crayons, or other small objects. Work with your child to show all the ways to make 10. (1 + 9, 6 + 4, and so on.) Together write addition sentences for each way to make ten.

Literature

Look for this book in a library. Ask your child to count how many objects are on each page.

How Many Snails?: A Counting Book
by Paul Giganti. Greenwillow Books, 1994

Carta para la casa

Querida familia:

Mi clase comenzó el Capítulo 1 esta semana. En este capítulo, aprenderé a sumar números hasta diez, cómo escribir oraciones numéricas de suma de distintas maneras y cómo usar ilustraciones para aprender a sumar mejor.

Con cariño, _____

Vocabulario

más (+) parte de una oración numérica de suma que significa "sumar a"

$$1 \overset{más}{+} 3 = 4$$

suma la respuesta a una oración numérica de suma

$$2 + 4 = ⑥$$

cero un número que significa nada; si sumas cero a cualquier número, ese número no cambia

$$1 + 0 = 1$$

Actividad para la casa

Use 10 fichas, crayolas u otros objetos pequeños. Trabaje con su hijo para mostrar todas las maneras de formar 10. ($1 + 9$, $6 + 4$ y así sucesivamente). Luego escriban juntos oraciones numéricas de suma para cada manera de formar diez.

Literatura

Busque este libro en una biblioteca. Pídale a su hijo que cuente cuántos objetos hay en cada página.

¿Cuántos caracoles?: Un libro para contar
por Paul Giganti.
Greenwillow Books, 1994

Name _____

Algebra • Use Pictures to Add To

Write how many.

1.

5 horses   and   3 more horses   ____ horses

2.

3 dogs   and   2 more dogs   ____ dogs

3.

4 cats   and   1 more cat   ____ cats

PROBLEM SOLVING REAL WORLD

There are 2 rabbits. 5 rabbits join
them. How many rabbits are there now?

There are ____ rabbits

Lesson Check

1. How many birds are there?

2 birds   and   6 more birds   _____ birds

4   6   8   9

○   ○   ◉   ○

Spiral Review

2. How many goats are there? (Lesson 1.1)

2 goats   and   4 more goats   _____ goats

2   6   8   10

○   ○   ○   ◉

3. How many rabbits are there? (Lesson 1.1)

5 rabbits   and   1 more rabbit   _____ rabbits

2   4   5   6

○   ○   ◉   ○

4. How many ducks are there? (Lesson 1.1)

5 ducks   and   4 more ducks   _____ ducks

9   8   5   1

◉   ○   ○   ○

Name _____

Model Adding To

Use to show adding to.
Draw the . Write the sum.

1. 5 ants and 1 more ant

$$5 + 1 = \underline{6}$$

2. 3 cats and 4 more cats

$$3 + 4 = \underline{P}$$

3. 4 dogs and 4 more dogs

$$4 + 4 = \underline{8}$$

4. 4 bees and 5 more bees

$$4 + 5 = \underline{10}$$

PROBLEM SOLVING **REAL WORLD**

Use the picture to help you complete the
addition sentences. Write each sum.

5. $\underline{5}$ ■ + $\bigcirc$ ■ = $\underline{5}$ ■ in all

6. $\underline{3}$ ● + $\underline{1}$ ● = $\underline{4}$ ● in all

Lesson Check

1. What is the sum of 4 and 2?

    2   4   6   8

    ○   ○   ○   ○

Spiral Review

2. How many butterflies are there? (Lesson 1.1)

   5 butterflies   and   2 more butterflies   _____ butterflies

    2   3   6   7

    ○   ○   ○   ○

3. What is the sum of 2 and 3? (Lesson 1.2)

    1   2   5   6

    ○   ○   ○   ○

4. How many birds are there? (Lesson 1.1)

   6 birds   and   1 more bird   _____ birds

    8   7   6   5

    ○   ○   ○   ○

Name _____

Model Putting Together

Use ○ to solve. Draw to show your work. Write the number sentence and how many.

1. There are 2 big dogs and 4 small dogs. How many dogs are there?

   ____ dogs

   ___ ○ ___ ○ ___

2. There are 3 red crayons and 2 green crayons. How many crayons are there?

   ____ crayons

   ___ ○ ___ ○ ___

3. There are 5 brown rocks and 3 white rocks. How many rocks are there?

   ____ rocks

   ___ ○ ___ ○ ___

PROBLEM SOLVING  REAL WORLD

4. Write your own addition story problem.

   _____
   - - - - - - - - - - - - - - - - -
   _____
   _____
   _____

Lesson Check

1. There are 3 black cats and 2 brown cats.
   How many cats are there?

   6   5   1   0
   ○   ○   ○   ○

2. There are 4 red flowers and 3 yellow flowers.
   How many flowers are there?

   1   3   7   8
   ○   ○   ○   ○

Spiral Review

3. How many turtles are there? (Lesson 1.1)

   6 turtles   and   3 more turtles   ____ turtles

   2   5   8   9
   ○   ○   ○   ○

4. What is the sum of 2 and 1? (Lesson 1.2)

   4   3   2   1
   ○   ○   ○   ○

Problem Solving • Model Addition

**Read the problem. Use the bar
model to solve. Complete the model
and the number sentence.**

1. Dylan has 7 flowers.
   4 of the flowers are red.
   The rest are yellow.
   How many flowers are yellow?

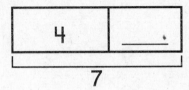

4 + _____ = 7

2. Some birds are flying in a group.
   4 more birds join the group.
   Then there are 9 birds in the
   group. How many birds were in
   the group before?

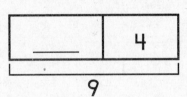

_____ + 4 = 9

3. 6 cats are walking.
   1 more cat walks with them.
   How many cats are walking now?

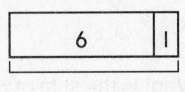

6 + 1 = _____

Lesson Check

1. 3 ducks are in the pond.
   6 more ducks join them.
   How many ducks are in the pond now?

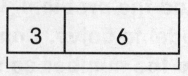

    9   6   3   1

    ○   ○   ○   ○

Spiral Review

2. There are 4 green grapes and 4 red grapes.
   How many grapes are there? (Lesson 1.3)

    1   5   8   9

    ○   ○   ○   ○

3. What is the sum of 7 and 3? (Lesson 1.2)

    10   9   7   4

    ○   ○   ○   ○

4. What is the sum of 6 and 2? (Lesson 1.2)

    9   8   7   4

    ○   ○   ○   ○

Algebra • Add Zero

Draw circles to show the number.
Write the sum.

1.

$$3 + 0 = \underline{\hspace{1cm}}$$

2.

$$0 + 5 = \underline{\hspace{1cm}}$$

3.

$$1 + 3 = \underline{\hspace{1cm}}$$

4.

$$5 + 1 = \underline{\hspace{1cm}}$$

PROBLEM SOLVING  REAL WORLD

Write the addition sentence to solve.

5. 6 turtles swim.
No turtles join them.
How many turtles are there now? $\underline{\hspace{0.8cm}} + \underline{\hspace{0.8cm}} = \underline{\hspace{0.8cm}}$

_____ turtles

Lesson Check

1. What is the sum for 0 + 4?

6	5	4	0
○	○	○	○

Spiral Review

2. There are 3 goats are in the barn.
4 more goats join them.
How many goats are in the barn now? **(Lesson 1.4)**

5	6	7	8
○	○	○	○

3. There are 7 blue crayons and 1 yellow crayon.
How many crayons are there? **(Lesson 1.3)**

4	6	7	8
○	○	○	○

4. What is the sum of 3 and 3? **(Lesson 1.2)**

3	4	5	6
○	○	○	○

Name _____

Algebra • Add in Any Order

Use . Write the sum.
Circle the addition sentences
in each row that have the same
addends in a different order.

1. $1 + 3 =$ ___ $1 + 2 =$ ___ $3 + 1 =$ ___

2. $2 + 3 =$ ___ $3 + 2 =$ ___ $0 + 5 =$ ___

3. $2 + 4 =$ ___ $3 + 3 =$ ___ $4 + 2 =$ ___

4. $4 + 1 =$ ___ $1 + 4 =$ ___ $0 + 4 =$ ___

5. $3 + 6 =$ ___ $4 + 5 =$ ___ $5 + 4 =$ ___

PROBLEM SOLVING REAL WORLD

Draw pictures to match the addition sentences.
Write the sums.

6. $5 + 2 =$ ___

   $2 + 5 =$ ___

Lesson Check

1. Which shows the same addends
   in a different order?

$$6 + 1 = 7$$

$1 + 5 = 6$	$1 + 6 = 7$	$2 + 6 = 8$	$7 + 1 = 8$
○	○	○	○

Spiral Review

2. What is the sum? (Lesson 1.5)

$$0 + 2 = \underline{\quad}$$

0	2	3	5
○	○	○	○

3. There are 5 long strings and 3 short strings.
   How many strings are there? (Lesson 1.3)

5	7	8	9
○	○	○	○

4. What is the sum of 6 and 2? (Lesson 1.2)

10	8	6	4
○	○	○	○

Lesson Check

1. Which shows a way to make 10?

$1 + 7$ ⃝ $2 + 6$ ⃝ $4 + 6$ ⃝ $5 + 4$ ⃝

2. Which shows a way to make 6?

 ⃝  ⃝  ⃝ 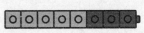 ⃝

Spiral Review

3. Which shows the same addends in a different order? (Lesson 1.6)

$$2 + 4 = 6$$

$1 + 5 = 6$ ⃝ | $4 + 2 = 6$ ⃝ | $6 + 2 = 8$ ⃝ | $4 + 1 = 5$ ⃝

4. What is the sum for $2 + 0$? (Lesson 1.5)

4 ⃝   3 ⃝   2 ⃝   1 ⃝

5. 3 rabbits sit in the grass.
   4 more rabbits join them.
   How many rabbits are there now? (Lesson 1.4)

4 ⃝   5 ⃝   6 ⃝   7 ⃝

Name _____

Algebra • Put Together Numbers to 10

Use 🎲 🎲. Color to show how to make 8. Complete the addition sentences.

1. $8 = \underline{8} + \underline{0}$

2. $8 = \underline{} + \underline{}$

3. $8 = \underline{} + \underline{}$

4. $8 = \underline{} + \underline{}$

5. $8 = \underline{} + \underline{}$

6. $8 = \underline{} + \underline{}$

7. $8 = \underline{} + \underline{}$

8. $8 = \underline{} + \underline{}$

9. $8 = \underline{} + \underline{}$

School-Home Letter

Dear Family,

My class started Chapter 2 this week. In this chapter, I will learn different ways to subtract. I will learn to write subtraction sentences.

Love, _____

Vocabulary

minus (−) part of a subtraction sentence that means "to take from"

minus
$6 - 5 = 1$

difference answer to a subtraction sentence

$3 - 2 = \text{①}$

fewer smaller number of something 3 books and 2 bags, you have 1 fewer bag than books

Home Activity

Show your child two groups of household objects, such as spoons and forks. Have your child use subtraction to compare how many more or fewer. Use different amounts and different objects every day.

$5 - 2 = ?$

Literature

Look for these books in a library. Have your child compare groups of items using *more* and *fewer*.

More, Fewer, Less by Tana Hoban. Greenwillow Books, 1998.

Elevator Magic by Stuart J. Murphy. HarperCollins, 1997.

Carta
para la casa

Querida familia:

Mi clase comenzó el Capítulo 2 esta semana. En este capítulo, aprenderé distintas formas para restar. Aprenderé a escribir enunciados de resta.

Con cariño, _____

Vocabulario

menos (−) parte de un enunciado de resta que significa "quitar de"

$$menos$$
$$6 - 5 = 1$$

diferencia respuesta a un enunciado de resta

$$3 - 2 = \textcircled{1}$$

menos un número Cantidad menor de algo. Si tienes 3 libros y 2 carteras, tienes 1 cartera menos.

Actividad para la casa

Muestre a su hijo dos grupos de objetos que haya en la casa, como cucharas y tenedores. Pídale que use la resta para comparar cuántos objetos más o menos hay de cada tipo. Use distintas cantidades y objetos diferentes cada día.

$$5 - 2 = ?$$

Literatura

Busque estos libros en una biblioteca. Pídale a su hijo que compare grupos de cosas usando *más* y *menos*.

More, Fewer, Less por Tana Hoban. Greenwillow Books, 1998.

El ascensor maravilloso por Stuart J. Murphy. HarperCollins, 1997.

Name _____

Use Pictures to Show Taking From

**Use the picture. Circle the part
you take from the whole group.
Then cross it out. Write how many
there are now.**

1.

3 cats I cat walks away. **2** cats now

2.

5 horses 2 horses walk away. **3** horses now

3.

7 dogs 3 dogs walk away. **4** dogs now

PROBLEM SOLVING REAL WORLD

Solve.

4. There are 7 birds. 2 birds fly away.
   How many birds are there now?

   5 birds

Lesson Check

1. There are 4 ducks.
   2 ducks swim away.
   How many ducks are
   there now?

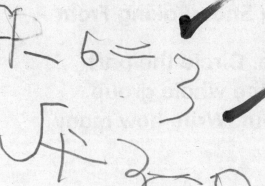

    6   5   3   2

    ○   ○   ○   ○

Spiral Review

2. What is the sum for 2 + 0? (Lesson 1.5)

    0   1   2   4

    ○   ○   ●   ○

3. How many birds? (Lesson 1.1)

5 birds   and   2 birds   _____ birds

    7   5   3   2

    ●   ○   ○   ○

4. What is the sum? (Lesson 1.8)

$$\begin{array}{r} 6 \\ + 2 \\ \hline \end{array}$$

    9   8   6   4

    ○   ●   ○   ○

Model Taking From

Use 🎲 to show taking from.
Draw the 🎲. Circle the part
you take from the group. Then
cross it out. Write the difference.

1. 4 turtles   1 turtle walks away.

$$4 - 1 = \underline{3}$$

2. 8 birds 7 birds fly away.

$$8 - 7 = \underline{1}$$

3. 6 bees   2 bees fly away.

$$6 - 2 = \underline{4}$$

4. 7 swans 5 swans swim away.

$$7 - 5 = \underline{1}$$

PROBLEM SOLVING REAL WORLD

Draw 🎲 to solve. Complete
the subtraction sentence.

5. There are 8 fish.
4 fish swim away.
How many fish
are there now?

$$\underline{8} - \underline{4} = \underline{1}$$

_____ fish

Lesson Check

1. What is the difference?

2	3	4	5
○	○	○	○

Spiral Review

2. How many snails? (Lesson 1.1)

7 snails   and   1 snail   ___ snails

5	6	7	8
○	○	○	○

3. Which shows the same addends in a different order? (Lesson 1.6)

$$6 + 2 = 8$$

$7 + 1 = 8$ $2 + 6 = 8$
   ○   ○

$8 - 6 = 2$ $8 - 2 = 6$
   ○   ○

Name _____

Model Taking Apart

Use ⬤ to solve. Draw to show your work. Write the number sentence and how many.

1. There are 7 bags. 2 bags are big. The rest are small. How many bags are small?

____ small bags   __ ◯ __ ◯ __

···

2. There are 6 dogs. 4 dogs are brown. The rest are black. How many dogs are black?

____ black dogs   __ ◯ __ ◯ __

PROBLEM SOLVING REAL WORLD

Solve. Draw a model to explain.

3. There are 8 cats. 6 cats walk away. How many cats are left?

____ cats left

Lesson Check

1. Which number sentence solves the problem?
There are 8 blocks. 3 blocks are white. The rest are blue. How many blocks are blue?

$3 + 3 = 6$ | $5 - 3 = 2$ | $8 - 3 = 5$ | $2 + 8 = 10$
○ | ○ | ○ | ○

Spiral Review

2. There are 4 green grapes and 5 red grapes. How many grapes are there? (Lesson 1.3)

10 9 8 5
○ ○ ○ ○

3. 3 ducks swim in the pond. 2 more join them. How many ducks are in the pond now? (Lesson 1.4)

1 2 3 5
○ ○ ○ ○

4. What is the sum of 1 and 4? (Lesson 1.2)

5 4 3 2
○ ○ ○ ○

Name _____

Problem Solving • Model Subtraction

**Read the problem. Use the model to solve.
Complete the model and the number sentence.**

1. There were 7 ducks in the pond. Some ducks swam away. Then there were 4 ducks. How many ducks swam away?

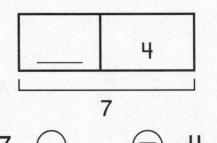

$7 \bigcirc \_\_\_ \bigcirc 4$

2. Tom had 9 gifts. He gave some away. Then there were 6 gifts. How many gifts did he give away?

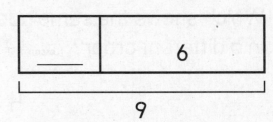

$9 \bigcirc \_\_\_ \bigcirc 6$

3. Some ponies were in a barn. 3 ponies walked out. Then there were 2 ponies. How many ponies were in the barn before?

$\_\_\_ \bigcirc 3 \bigcirc 2$

4. There are 10 puppies. 3 puppies are brown. The rest are black. How many puppies are black?

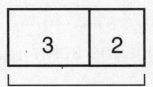

$10 \bigcirc 3 \bigcirc \_\_\_$

Lesson Check

1. There are 8 shells. 6 shells are white. The rest are pink. How many shells are pink?

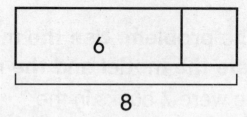

8   6   3   2

○   ○   ○   ○

Spiral Review

2. Which shows the same addends in a different order? (Lesson 1.6)

$$4 + 2 = 6$$

$6 - 2 = 4$ $2 + 4 = 6$ $2 + 2 = 4$ $6 - 4 = 2$

○   ○   ○   ○

3. What is the sum? (Lesson 1.8)

$$\begin{array}{r} 4 \\ + 3 \\ \hline \end{array}$$

1   6   7   8

○   ○   ○   ○

Use Pictures and Subtraction to Compare

Draw lines to match.
Subtract to compare.

1.

8 − 5 = ____   ____ more

2.

9 − 4 = ____   ____ fewer

PROBLEM SOLVING

Draw a picture to show the problem.
Write a subtraction sentence to
match your picture.

3. Jo has 4 golf clubs and
   2 golf balls. How many fewer
   golf balls does Jo have?

____ − ____ = ____   ____ fewer

Lesson Check

1. How many fewer 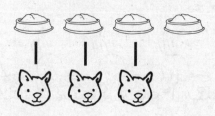 are there?

4	3	1	0
○	○	○	○

Spiral Review

2. What is the sum of 5 and 1? (Lesson 1.2)

1	4	5	6
○	○	○	○

3. What is the sum? (Lesson 1.8)

$$\begin{array}{r} 4 \\ + 5 \\ \hline \end{array}$$

9	8	5	4
○	○	○	○

Name _____

Subtract to Compare

Read the problem. Use the bar model to solve. Write the number sentence. Then write how many.

1. Ben has 7 flowers. Tim has 5 flowers. How many fewer flowers does Tim have than Ben?

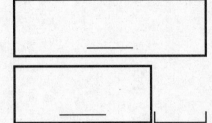

_____ fewer flowers

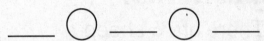

2. Nicky has 8 toys. Ada has 3 toys. How many more toys does Nicky have than Ada?

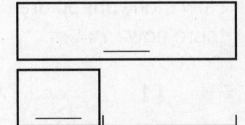

_____ more toys

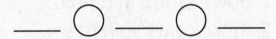

PROBLEM SOLVING REAL WORLD

Complete the number sentence to solve.

3. Maya has 7 pens. Sam has 1 pen. How many more pens does Maya have than Sam?

___ – ___ = ___

 more pens

Lesson Check

1. Jesse has 2 stickers.
   Sara has 8 stickers. How
   many fewer stickers does
   Jesse have than Sara?

8

2

   2   4   6   8
   ○   ○   ○   ○

Spiral Review

2. There are 6 sheep.
   5 sheep walk away.
   How many sheep are
   there now? (Lesson 2.1)

   11   6   5   1
   ○   ○   ○   ○

3. 5 cows stand in a field.
   2 more cows join them.
   How many cows are in
   the field now? (Lesson 1.4)

   7   5   3   2
   ○   ○   ○   ○

Subtract All or Zero

Complete the subtraction sentence.

1.

   $3 - 0 =$ ___

2.

   $2 - 2 =$ ___

3. $5 - 0 =$ ___

4. ___ $= 1 - 0$

5. $6 - 6 =$ ___

6. $0 =$ ___ $- 8$

7. $7 - 0 =$ ___

8. $5 -$ ___ $= 0$

PROBLEM SOLVING

Write the number sentence
and tell how many.

9. There are 9 books on the shelf.
   9 are blue and the rest are green.
   How many books are green?

   ___ green books

Lesson Check

1. What is the difference for 4 − 0?

0 ⭘　　2 ⭘　　3 ⭘　　4 ⭘

2. What is the difference for 6 − 6?

0 ⭘　　1 ⭘　　5 ⭘　　6 ⭘

Spiral Review

3. How many bunnies? (Lesson 1.1)

3 bunnies　　and　　3 bunnies　　____ bunnies

0 ⭘　　1 ⭘　　6 ⭘　　7 ⭘

4. Which shows a way to make 9? (Lesson 1.7)

⭘

⭘

⭘

⭘

Name _____

Algebra • Take Apart Numbers

Use [🎲]. Color and draw to show
how to take apart 5. Complete the
subtraction sentence.

1. ⬜⬜⬜⬜⬜   5 − ___ = ___

2. ⬜⬜⬜⬜⬜   5 − ___ = ___

3. ⬜⬜⬜⬜⬜   5 − ___ = ___

4. ⬜⬜⬜⬜⬜   5 − ___ = ___

5. ⬜⬜⬜⬜⬜   5 − ___ = ___

6. ⬜⬜⬜⬜⬜   5 − ___ = ___

PROBLEM SOLVING

Solve.

7. Joe has 9 marbles. He gives
   them all to his sister. How many
   marbles does he have now?

   _____ marbles

Lesson Check

1. Which shows a way to take apart 8?

$$9 - 9 = 0 \quad | \quad 9 - 8 = 1 \quad | \quad 8 - 1 = 7 \quad | \quad 8 + 8 = 16$$

○　　　　　○　　　　　○　　　　　○

Spiral Review

2. What is the sum? (Lesson 1.8)

$$\begin{array}{r} 6 \\ + 4 \\ \hline \end{array}$$

10　　　　9　　　　6　　　　2
○　　　　○　　　　○　　　　○

3. There are 7 fish.
3 fish swim away.
How many fish are there now? (Lesson 2.1)

2　　　　3　　　　4　　　　10
○　　　　○　　　　○　　　　○

Subraction from 10 or Less

Write the difference.

1. 5
   −1

2. 3
   −2

3. 8
   −3

4. 6
   −4

5. 7
   −0

6. 5
   −3

7. 4
   −4

8. 8
   −1

9. 8
   −7

10. 6
    −3

11. 5
    −5

12. 7
    −6

PROBLEM SOLVING

Solve.

13. 6 birds are in the tree.
    None of the birds fly away.
    How many birds are left?

    ____ − ____ = ____

Lesson Check

1. What is the difference?

$$\begin{array}{r} 4 \\ -\ 0 \\ \hline \end{array}$$

10	9	4	0
○	○	○	○

Spiral Review

2. Which number sentence solves
the problem?
There are 8 pens. 3 pens
are blue. The rest are red.
How many pens are red? (Lesson 2.3)

$8 + 3 = 11$	$3 - 1 = 2$	$8 + 1 = 9$	$8 - 3 = 5$
○	○	○	○

3. Which shows the same
addends in a different order? (Lesson 1.6)

$$5 + 4 = 9$$

$4 + 5 = 9$	$4 + 4 = 8$	$5 + 3 = 8$	$9 - 4 = 5$
○	○	○	○

Chapter 2 Extra Practice

Lessons 2.1 - 2.2 (pp. 53 – 60) ·

Use 🎲 to show taking from. Draw the 🎲.
Circle the part you take from the group.
Then cross it out. Write the difference.

1. 5 whales   3 whales swim away.

$$5 - 3 = \underline{\quad}$$

Lesson 2.3 (pp. 61 – 64) ·

Use ⬤ to solve. Draw to show your work.
Write the number sentence and how many.

1. There are 7 snails. 2 snails
   are big. The rest are small.
   How many snails are small?

___ small snails   ___ ◯ ___ ◯ ___

Lessons 2.5 - 2.6 (pp. 69 – 75) ·

Read the problem. Use the bar model
to solve. Write the number sentence.
Then write how many.

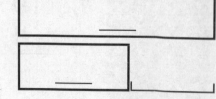

1. Tony has 9 buckets. Gina has 5 buckets.
   How many fewer buckets does Gina
   have than Tony?

___ fewer buckets   ___ ◯ ___ ◯ ___

plete the subtraction sentence.

___ − 0 = ___

2. $4 -$ ___ $= 0$

___ $= 8 - 8$

4. $9 - 0 =$ ___

Lesson 2.8 (pp. 81 – 84)

Use ⬚. Color and draw to show how to take apart 7. Complete the subtraction sentence.

I. ⬚⬚⬚⬚⬚⬚⬚

$7 -$ ___ $=$ ___

2. ⬚⬚⬚⬚⬚⬚⬚

$7 -$ ___ $=$ ___

Lesson 2.9 (pp. 85 – 88)

Write the difference.

I. $\begin{array}{r} 8 \\ -\ 8 \\ \hline \end{array}$

2. $\begin{array}{r} 5 \\ -\ 4 \\ \hline \end{array}$

3. $\begin{array}{r} 10 \\ -\ 0 \\ \hline \end{array}$

4. $\begin{array}{r} 9 \\ -\ 4 \\ \hline \end{array}$

5. $\begin{array}{r} 7 \\ -\ 6 \\ \hline \end{array}$

6. $\begin{array}{r} 9 \\ -\ 2 \\ \hline \end{array}$

7. $\begin{array}{r} 10 \\ -\ 4 \\ \hline \end{array}$

8. $\begin{array}{r} 5 \\ -\ 2 \\ \hline \end{array}$

9. $\begin{array}{r} 6 \\ -\ 1 \\ \hline \end{array}$

10. $\begin{array}{r} 8 \\ -\ 6 \\ \hline \end{array}$

School-Home Letter

Dear Family,

My class started Chapter 3 this week. In this chapter, I will learn about addition strategies such as counting on, adding doubles, and adding in any order.

Love, _____

Vocabulary

doubles Two equal groups make a doubles fact.

$2 + 2 = 4$

doubles plus one $2 + 2 = 4$, so $2 + 3$ is 1 more, or 5.

$2 + 3 = 5$

Home Activity

Have your child find objects that show doubles facts. For example, he or she may find a pair of shoes to show $1 + 1$, a carton of eggs to show $6 + 6$, or a six-pack of juice cans to show $3 + 3$. Ask your child to say those doubles facts.

Add another item to one of the groups, and have your child name the doubles plus one fact.

Literature

Look for these books in a library. Point out examples of doubles and counting on facts in the pictures.

12 Ways to Get to 11 by Eve Merriam. Aladdin, 1996.

Two of Everything: A Chinese Folktale by Lily Toy Hong. Albert Whitman and Company, 1993.

Carta
para la casa

Querida familia:

Mi clase comenzó el Capítulo 3 esta semana. En este capítulo, aprenderé estrategias de suma como contar hacia adelante, sumar dobles y sumar en cualquier orden.

Con cariño, _____

Vocabulario

dobles Dos grupos iguales forman una operación de dobles.

$$2 + 2 = 4$$

dobles más uno $2 + 2 = 4$, por lo tanto $2 + 3$ es 1 más o 5.

$$2 + 3 = 5$$

Actividad para la casa

Pida a su hijo que encuentre objetos que muestren operaciones de dobles. Por ejemplo, puede hallar un par de zapatos para mostrar $1 + 1$, una caja de huevos para mostrar $6 + 6$ o una caja de 6 latas de jugo para mostrar $3 + 3$. Pídale a su hijo que le diga cuáles son las operaciones de dobles.

Agregue otro artículo a uno de los grupos y pida a su hijo que nombre la operación de dobles más uno.

Literatura

Busque estos libros en una biblioteca. Señale ejemplos de operaciones de dobles y de contar uno hacia delante en las imágenes.

12 Ways to Get to 11
por Eve Merriam. Aladdin, 1996.

Two of Everything: A Chinese Folktale
por Lily Toy Hong. Albert Whitman and Company, 1993.

Add Doubles

Use ▪. Draw ▪ to show your work.
Write the sum.

1.   4
   + 4
      ‾‾‾

2.   6
   + 6
      ‾‾‾

3.   3
   + 3
      ‾‾‾

4.   8
   + 8
      ‾‾‾

5.   5
   + 5
      ‾‾‾

6.   7
   + 7
      ‾‾‾

PROBLEM SOLVING REAL WORLD

Write a doubles fact to solve.

7. There are 16 crayons in the box.
Some are green and some are red.
The number of green crayons is the
same as the number of red crayons.

____ = ____ + ____

Lesson Check

1. Which is a doubles fact?

$8 + 3 = 11$
◯

$9 + 9 = 18$
◯

$1 + 5 = 6$
◯

$5 + 7 = 12$
◯

2. Which is a doubles fact?

$6 + 3 = 9$
◯

$6 + 6 = 12$
◯

$6 + 4 = 10$
◯

$6 + 7 = 13$
◯

Spiral Review

3. What is the sum of 3 and 2? (Lesson 1.2)

7   6   5   1
◯   ◯   ◯   ◯

4. What is the sum for $4 + 0$? (Lesson 1.5)

5   4   3   0
◯   ◯   ◯   ◯

Name _____

Use Doubles to Add

Use 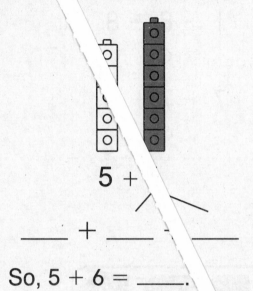. Make doubles. Add.

1.

$5 +$ ___

___ + ___ + ___

So, $5 + 6 =$ ___.

2.

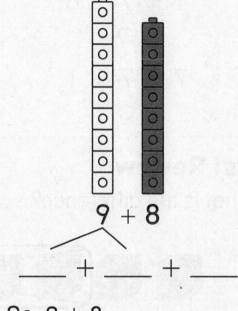

$9 + 8$

___ + ___ + ___

So, $9 + 8 =$ ___.

Use doubles to help you add.

3. $8 + 7 =$ ___

4. $6 + 5 =$ ___

5. $7 + 6 =$ ___

6. $4 + 5 =$ ___

7. $7 + 8 =$ ___

8. $8 + 9 =$ ___

PROBLEM SOLVING REAL WORLD

Solve. Draw or write to explain.

9. Bo has 6 toys. Mia has 7 toys.
   How many toys do they have?

   ____ toys

Lesson Check

1. Which has the same sum as 7 + 8?

1 + 7 + 8
○

1 + 8 + 8
○

7 + 7 + 1
○

7 + 7 + 2
○

Spiral Review

2. What is the difference? (Lesson 2.2)

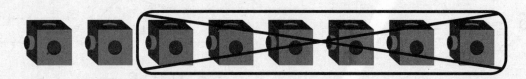

0
○

2
○

6
○

8
○

3. There are 7 gray kittens.
2 kittens are black. How many
fewer black kittens are there
than gray kittens? (Lesson 2.6)

7

2.

?

9
○

7
○

5
○

2
○

Name _____

Add 10 and More

Draw red ○ to show 10. Draw
yellow ○ to show the other addend.
Write the sum.

1.
$$10$$
$$+ \ 7$$

2.
$$10$$
$$+ \ 5$$

3.
$$10$$
$$+ \ 9$$

4.
$$10$$
$$+ \ 4$$

PROBLEM SOLVING REAL WORLD

Draw red and yellow ○ to solve.
Write the addition sentence.

5. Linda has 10 toy cars.
   She gets 6 more cars.
   How many toy cars
   does she have now?

_____ + _____ = _____ toy cars

Lesson Check

1. How many would you need to show the addition fact?

$$\begin{array}{r} 10 \\ + 3 \\ \hline \end{array}$$

13	7	3	2
○	○	○	○

2. What number sentence does this model show?

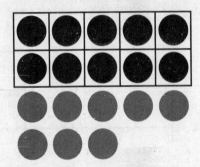

- ○ $8 + 8 = 16$
- ○ $9 + 6 = 15$
- ○ $10 + 7 = 17$
- ○ $10 + 8 = 18$

Spiral Review

3. Which shows a way to make 10? (Lesson 1.7)

$5 + 4$	$6 + 4$	$7 + 2$	$8 + 1$
○	○	○	○

4. There are 3 large turtles and 1 small turtle. How many turtles are there? (Lesson 1.3)

2	4	6	8
○	○	○	○

Make a 10 to Add

Use red and yellow ⬡ and a ten
frame. Show both addends. Draw to make
a ten. Then write the new fact. Add.

1. $\begin{array}{r} 5 \\ + 7 \\ \hline \end{array}$ 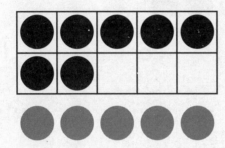 $\begin{array}{c} \square \\ + \\ \square \\ \hline \square \end{array}$

2. $\begin{array}{r} 9 \\ + 5 \\ \hline \end{array}$ $\begin{array}{c} \square \\ + \\ \square \\ \hline \square \end{array}$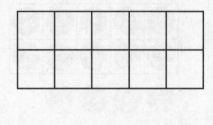

3. $\begin{array}{r} 8 \\ + 3 \\ \hline \end{array}$ $\begin{array}{c} \square \\ + \\ \square \\ \hline \square \end{array}$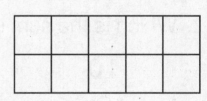

PROBLEM SOLVING

Solve.

4. $10 + 6$ has the same sum as $7 +$ _____.

Lesson Check

1. What sum does this model show?

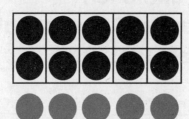

5	10	15	16
○	○	○	○

2. What addition sentence does this model show?

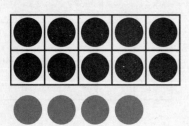

○ $8 + 5 = 13$

○ $9 + 4 = 13$

○ $10 + 3 = 13$

○ $10 + 4 = 14$

Spiral Review

3. Which is the sum of $4 + 6$? (Lesson 1.8)

10	9	3	2
○	○	○	○

4. There are 2 big flowers and 4 small flowers. How many flowers are there? (Lesson 1.3)

5	6	13	14
○	○	○	○

Use Make a 10 to Add

**Write to show how you make a ten.
Then add.**

1. What is $9 + 7$?

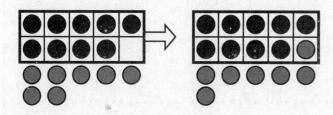

___ + ___ + ___

___ + ___ = ___

So, $9 + 7 =$ ___.

2. What is $5 + 8$?

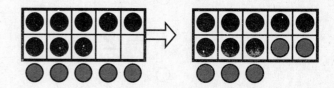

___ + ___ + ___

___ + ___ = ___

So, $5 + 8 =$ ___.

PROBLEM SOLVING REAL WORLD

Use the clues to solve.
Draw lines to match.

3. Ann and Gia are eating grapes.
Ann eats 10 green grapes and
6 red grapes. Gia eats the same
number of grapes as Ann. Match
each person to her grapes.

Ann	7 green grapes and 9 red grapes
Gia	10 green grapes and 6 red grapes

Lesson Check

1. Which shows how to make a ten to find 8 + 4?

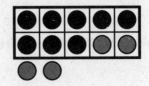

$8 + 2 + 3$
○

$8 + 2 + 2$
○

$8 + 1 + 2$
○

$5 + 3 + 2$
○

Spiral Review

2. What is the difference? (Lesson 2.7)

$$9 - 9 = \underline{\quad}$$

0
○

1
○

9
○

18
○

3. What is the difference? (Lesson 2.9)

$$\begin{array}{r} 8 \\ -2 \\ \hline \end{array}$$

10
○

6
○

5
○

4
○

Algebra • Add 3 Numbers

Look at the . Complete the addition sentences showing two ways to find the sum.

1. 5 + 4 + 2 = ____

___ + ___ = ___   ___ + ___ = ___

2. 2 + 2 + 6 = ____

___ + ___ = ___   ___ + ___ = ___

PROBLEM SOLVING

3. Choose three numbers from 1 to 6.
   Write the numbers in an addition sentence.
   Show two ways to find the sum.

Lesson Check

1. What is the sum of $3 + 4 + 2$?

11	10	9	6
○	○	○	○

2. What is the sum of $5 + 1 + 4$?

0	10	11	12
○	○	○	○

Spiral Review

3. What is the sum? **(Lesson 1.8)**

$$3 + 7 = \underline{\quad}$$

3	4	9	10
○	○	○	○

4. 4 cows are in the barn. 2 more cows join them. How many cows are in the barn now? **(Lesson 1.4)**

2	6	7	8
○	○	○	○

© Houghton Mifflin Harcourt Publishing Company

Algebra • Add 3 Numbers

Choose a strategy.
Circle two addends to add first.
Write the sum.

1.
```
  7
  3
+ 3
───
```

2.
```
  2
  2
+ 6
───
```

3.
```
  6
  6
+ 3
───
```

4.
```
  2
  0
+ 8
───
```

5.
```
  1
  2
+ 9
───
```

6.
```
  6
  4
+ 3
───
```

7.
```
  3
  3
+ 5
───
```

8.
```
  4
  4
+ 8
───
```

PROBLEM SOLVING  REAL WORLD

Draw a picture. Write the number sentence.

9. Don has 4 black dogs.
   Tim has 3 small dogs.
   Sue has 3 big dogs.
   How many dogs do they have?

   ___ + ___ + ___ = ___ dogs

Lesson Check

1. What is the sum of 4 + 4 + 2?

4 ○   8 ○   10 ○   14 ○

2. What is the sum?

$$\begin{array}{r} 7 \\ 3 \\ +\ 2 \\ \hline \end{array}$$

5 ○   10 ○   11 ○   12 ○

Spiral Review

3. Which is a doubles plus one fact? (Lesson 3.5)

1 + 1 = 2 | 4 + 2 = 6 | 3 + 4 = 7 | 5 + 3 = 8
○   ○   ○   ○

4. What number sentence does this model show? (Lesson 3.8)

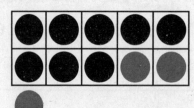

10 + 3 = 13 | 10 + 1 = 11 | 6 + 5 = 11 | 10 + 4 = 14
○   ○   ○   ○

Name _____

Problem Solving • Use Addition Strategies

Draw a picture to solve.

1. Franco has 5 crayons. He gets 8 more crayons. Then he gets 2 more crayons. How many crayons does he have now?

___ ◯ ___ ◯ ___ ◯ ___   ___ crayons

2. Jackson has 6 blocks. He gets 5 more blocks. Then he gets 3 more blocks. How many blocks does he have now?

___ ◯ ___ ◯ ___ ◯ ___   ___ blocks

3. Avni has 7 gifts. Then he gets 2 more gifts. Then he gets 3 more gifts. How many gifts does Avni have now?

___ ◯ ___ ◯ ___ ◯ ___   ___ gifts

4. Meeka has 4 rings. She gets 2 more rings. Then she gets 1 more ring. How many rings does she have now?

___ ◯ ___ ◯ ___ ◯ ___   ___ rings

Lesson Check

1. Lila has 3 gray stones.
   She has 4 black stones.
   She also has 7 white stones.
   How many stones does she have?

7	10	13	14
○	○	○	○

2. Patrick has 3 red stickers, 6 pink
   stickers, and 8 green stickers. How
   many stickers does Patrick have?

18	17	16	14
○	○	○	○

Spiral Review

3. What is the sum of 2 + 4 or 4 + 2? **(Lesson 1.6)**

6	5	4	3
○	○	○	○

4. There are 6 black pens.
   There are 3 blue pens.
   How many pens are there? **(Lesson 1.3)**

2	5	8	9
○	○	○	○

Chapter 3 Extra Practice

Lesson 3.1 ....

Add. Change the order of the addends. Add again.

1. 9
   + 6
   _____

2. 7
   + 2
   _____

3. 0
   + 8
   _____

Lesson 3.2 ....

Circle the greater addend.
Count on to find the sum.

1. $7 + 2 = $ _____

2. $3 + 5 = $ _____

3. $4 + 3 = $ _____

Lesson 3.4 ....

Use 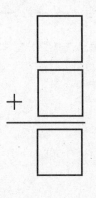 . Make doubles. Add.

1.
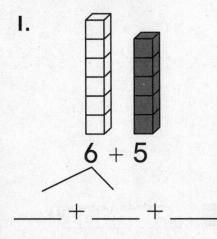

6 + 5

___ + ___ + ___

So, $6 + 5 = $ _____ .

2.
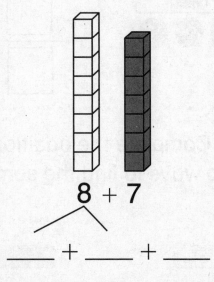

8 + 7

___ + ___ + ___

So, $8 + 7 = $ _____ .

3.
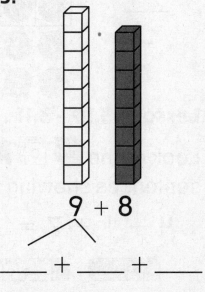

9 + 8

___ + ___ + ___

So, $9 + 8 = $ _____ .

Lesson 3.5 . . . .

Add. Write the doubles fact
you used to solve the problem.

1. $6 + 7 =$ ___

___ ◯ ___ ◯ ___

2. $5 + 6 =$ ___

___ ◯ ___ ◯ ___

Lesson 3.6 . . . .

Add. Color doubles facts 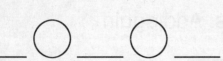.
Color count on facts .
Color doubles plus one or
doubles minus one facts .

1. $8 + 8 =$ ___

2. $5 + 6 =$ ___

3. $3 + 1 =$ ___

Lessons 3.7 – 3.8 . . . .

Use red and yellow and a ten frame. Show both addends.
Draw to make a ten. Then write the new fact. Add.

1. $\begin{array}{r} 9 \\ +\ 7 \\ \hline \end{array}$

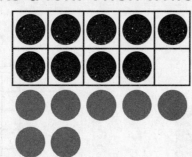

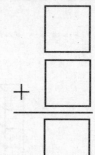

Lessons 3.10 – 3.11 . . . .

Look at the 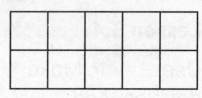. Complete the addition
sentences showing two ways to find the sum.

1. $4 + 1 + 7 =$ ___

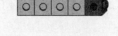

___ + ___ = ___   ___ + ___ = ___

School-Home Letter

Dear Family,

My class started Chapter 4 this week. In this chapter, I will learn about subtraction strategies and how to solve subtraction word problems.

Love, _____

Vocabulary

count back a way to subtract by counting back from the larger number

$$8 - 1 = 7$$

Start at 8.
Count back 1.
You are on 7.

Home Activity

Have your child practice counting from 1 to 8 and then from 8 to 1. Display numbers 1–8 on a piece of poster board or notebook paper. Each day, work with your child to solve simple subtraction problems by counting back 1, 2, or 3 using the list of numbers.

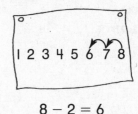

$$8 - 2 = 6$$

Literature

Look for these books in a library. Reading them together will reinforce your child's learning.

Monster Musical Chairs by Stuart J. Murphy. HarperCollins Children's Books, 2000.

Ten Little Ladybugs by Melanie Gerth. Piggy Toes Press, 2001.

Carta para la casa

Querida familia:

Mi clase comenzó hoy el Capítulo 4. En este capítulo, aprenderé estrategias de resta y a resolver problemas de resta en palabras.

Con cariño, _____

Vocabulario

contar hacia atrás un modo de restar contando hacia atrás de un número mayor

$$8 - 1 = 7$$

Comienza en 8.
Cuenta hacia atrás 1.
Quedas en 7.

Actividad para la casa

Pida a su hijo que cuente de 1 a 8 y de 8 a 1. Muestre los números de 1 a 8 en una cartulina o una hoja de cuaderno. Cada día, practique con su hijo resolver problemas simples de resta contando hacia atrás 1, 2 ó 3, en la lista de los números anotados.

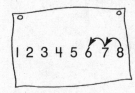

$$8 - 2 = 6$$

Literatura

Busquen estos libros en la biblioteca. Si los lee con su hijo, ayudará a reforzar su aprendizaje.

Monster Musical Chairs
Stuart J. Murphy. HarperCollins Children's Books, 2000.

Ten Little Ladybugs
Melanie Gerth. Piggy Toes Press, 2001.

Name _____

Think Addition to Subtract

Use  to add and to subtract.

1.
$$\begin{array}{r} 9 \\ - 3 \\ \hline ? \end{array}$$

Think
$$\begin{array}{r} 3 \\ + \boxed{} \\ \hline 9 \end{array}$$

So
$$\begin{array}{r} 9 \\ - 3 \\ \hline \end{array}$$

2.
$$\begin{array}{r} 15 \\ - 8 \\ \hline ? \end{array}$$

Think
$$\begin{array}{r} 8 \\ + \boxed{} \\ \hline 15 \end{array}$$

So
$$\begin{array}{r} 15 \\ - 8 \\ \hline \end{array}$$

3.
$$\begin{array}{r} 11 \\ - 7 \\ \hline ? \end{array}$$

Think
$$\begin{array}{r} 7 \\ + \boxed{} \\ \hline 11 \end{array}$$

So
$$\begin{array}{r} 11 \\ - 7 \\ \hline \end{array}$$

4.
$$\begin{array}{r} 13 \\ - 4 \\ \hline ? \end{array}$$

Think
$$\begin{array}{r} 4 \\ + \boxed{} \\ \hline 13 \end{array}$$

So
$$\begin{array}{r} 13 \\ - 4 \\ \hline \end{array}$$

5.
$$\begin{array}{r} 14 \\ - 6 \\ \hline ? \end{array}$$

Think
$$\begin{array}{r} 6 \\ + \boxed{} \\ \hline 14 \end{array}$$

So
$$\begin{array}{r} 14 \\ - 6 \\ \hline \end{array}$$

PROBLEM SOLVING

6. Write a number sentence to solve.
   I have 18 pieces of fruit.
   9 are apples.
   The rest are oranges.
   How many are oranges?

   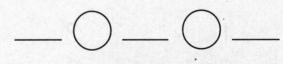

   ____ oranges

Lesson Check

1. Use the sum of $7 + 9$ to solve $16 - 9$.

     6   7   8   9

     ○   ○   ○   ○

2. What is the missing number?

$$5 + \boxed{} = 14 \qquad 14 - 5 = \boxed{}$$

     4   5   9   10

     ○   ○   ○   ○

Spiral Review

3. What is the sum? (Lesson 3.10)

$$4 + 4 + 6 = \underline{}$$

     10   14   15   16

     ○   ○   ○   ○

4. There are 4 birds.
3 birds fly away.
How many birds
are there now? (Lesson 2.1)

     1   3   4   7

     ○   ○   ○   ○

Name _____

Use Think Addition to Subtract

**Think of an addition fact
to help you subtract.**

1.   13
   − 8

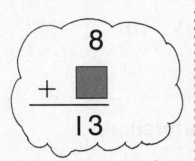

8
+ ▪
___
13

2.   12
   − 6

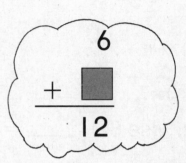

6
+ ▪
___
12

3.   6
  − 4

4.   14
  − 9

5.   9
  − 5

6.   13
  − 6

7.   10
  − 7

8.   12
  − 4

9.   16
  − 7

10.   11
   − 8

11.   14
   − 8

12.   15
   − 7

PROBLEM SOLVING  REAL WORLD

13. Solve. Draw or write to show your work.
    I have 15 nickels.
    Some are old. 6 are new.
    How many nickels are old?

    ____ nickels

Lesson Check

1. Use $9 + \underline{\quad} = 13$ to find the difference.

$$13 - 9 = \underline{\quad}$$

0	2	4	6
○	○	○	○

2. Use $8 + \underline{\quad} = 11$ to find the difference.

$$11 - 8 = \underline{\quad}$$

3	4	5	18
○	○	○	○

Spiral Review

3. What is the sum of this doubles plus one fact? (Lesson 3.5)

$$4 + 5 = \underline{\quad}$$

7	8	9	10
○	○	○	○

4. Count on to solve $7 + 2$. (Lesson 3.2)

4	5	8	9
○	○	○	○

Use 10 to Subtract

Use 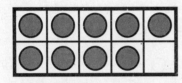 and ten frames. Make a
ten to subtract.
Draw to show your work.

1.

$12 - 9 = \underline{\quad?\quad}$

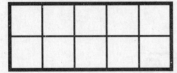

$12 - 9 = \underline{\quad\quad}$

2.

$12 - 8 = \underline{\quad?\quad}$

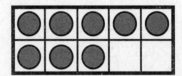

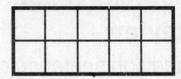

$12 - 8 = \underline{\quad\quad}$

PROBLEM SOLVING REAL WORLD

Solve. Use the ten frames to
make a ten to help you subtract.

3. Marta has 15 stickers.
8 are blue and the rest are red.
How many stickers are red?

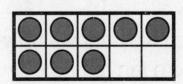

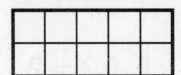

_____ stickers

Lesson Check

1. Which subtraction sentence do the ten frames show?

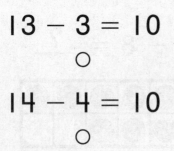

$13 - 3 = 10$
○

$13 - 9 = 4$
○

$14 - 4 = 10$
○

$14 - 5 = 9$
○

Spiral Review

2. What addition sentence does this model show? (Lesson 3.8)

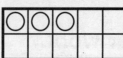
○

$3 + 2 = 5$
○

$5 + 2 = 7$
○

$10 + 1 = 11$
○

$11 + 1 = 12$
○

3. Which way shows how to make a ten to find $8 + 5$?

(Lesson 3.9)

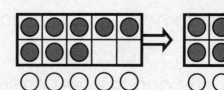
○ ○ ○ ○ ○   ○ ○ ○

$5 + 3 + 2$
○

$8 + 2 + 3$
○

$10 + 2 + 3$
○

$10 + 3 + 2$
○

Break Apart to Subtract

Subtract.

1. What is 13 − 5?

 Step 1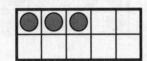

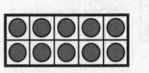 Step 2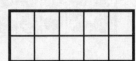

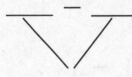

So, 13 − 5 = ___.

PROBLEM SOLVING REAL WORLD

Use the ten frames. Write a number sentence.

2. There are 17 goats in the barn. 8 goats go outside.
How many goats are still in the barn?

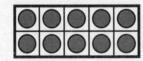 Step 1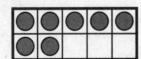

 Step 2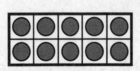

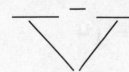

So, ___ − ___ = ___.

Lesson Check

1. Which way shows how to make a ten to find 12 − 4?

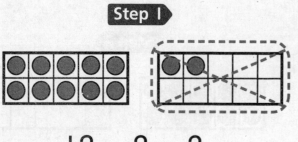 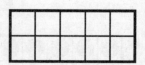

$12 - 2 - 2$
○

$10 - 4$
○

$12 - 1 - 1$
○

$10 - 1$
○

Spiral Review

2. Which shows a way to take apart 7? (Lesson 2.8)

$8 - 7 = 1$
○

$7 - 1 = 6$
○

$6 + 1 = 7$
○

$1 + 6 = 7$
○

3. Which is a doubles minus one fact? (Lesson 3.6)

$7 + 7 = 14$
○

$7 + 5 = 12$
○

$8 + 7 = 15$
○

$8 + 8 = 16$
○

Problem Solving • Use Subtraction Strategies

Act it out to solve.
Draw to show your work.

1. There are 13 monkeys.
   6 are small. The rest are big.
   How many monkeys are big?

   $13 - 6 = \boxed{}$

   _____ monkeys are big.

2. Mindy had 13 flowers. She
   gave some to Sarah. She
   has 9 left. How many flowers
   did she give to Sarah?

   $13 - \boxed{} = 9$

   Mindy gave _____ flowers to Sarah.

3. There are 5 more horses
   in the barn than outside.
   12 horses are in the barn.
   How many horses are
   outside?

   $12 - 5 = \boxed{}$

   _____ horses are outside.

4. Kim has 15 pennies.
   John has 6 pennies.
   How many fewer pennies
   does John have than Kim?

   $15 - 6 = \boxed{}$

   John has _____ fewer pennies.

Lesson Check

1. Jack has 14 oranges.
   He gives some away.
   He has 6 left.
   How many oranges did he give away?

   3   4   7   8
   ○   ○   ○   ○

2. 13 pears are in a basket.
   Some are yellow and some
   are green. 5 pears are green.
   How many pears are yellow?

   6   7   8   9
   ○   ○   ○   ○

Spiral Review

3. Rita has 4 plants.
   She gets 9 more plants.
   Then Rita gets 1 more plant.
   How many plants does she have now? (Lesson 3.12)

   14   13   10   5
   ○   ○   ○   ○

4. What is the sum of 10 + 5? (Lesson 3.7)

   15   10   5   4
   ○   ○   ○   ○

Chapter 4 Extra Practice

Lesson 4.1 (pp. 153–156) ·

Count back 1, 2, or 3.
Write the difference.

1. $10 - 3 =$ _____

2. _____ $= 11 - 3$

3. $9 - 2 =$ _____

4. _____ $= 6 - 1$

5. _____ $= 4 - 2$

6. $8 - 3 =$ _____

Lesson 4.2 – 4.3 (pp. 157-163) ·

Think of an addition fact
to help you subtract.

1.
$$\begin{array}{r} 13 \\ -\ 9 \\ \hline \end{array}$$

$$\begin{array}{r} 9 \\ +\ \blacksquare \\ \hline 13 \end{array}$$

2.
$$\begin{array}{r} 14 \\ -\ 7 \\ \hline \end{array}$$

$$\begin{array}{r} 7 \\ +\ \blacksquare \\ \hline 14 \end{array}$$

3.
$$\begin{array}{r} 14 \\ -\ 5 \\ \hline \end{array}$$

4.
$$\begin{array}{r} 5 \\ -\ 4 \\ \hline \end{array}$$

5.
$$\begin{array}{r} 12 \\ -\ 9 \\ \hline \end{array}$$

6.
$$\begin{array}{r} 13 \\ -\ 6 \\ \hline \end{array}$$

7.
$$\begin{array}{r} 15 \\ -\ 8 \\ \hline \end{array}$$

Lesson 4.4 (pp.165–168) ·

Use ● and ten frames.
Make a ten to subtract.
Draw to show your work.

I.

$16 - 9 =$ ___?___

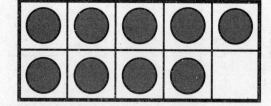

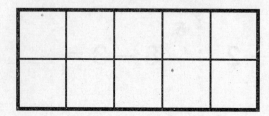

$16 - 9 =$ _____

2.

$13 - 8 =$ ___?___

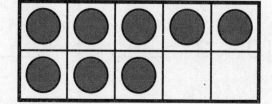

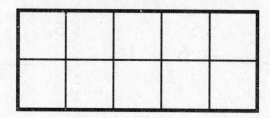

$13 - 8 =$ _____

Lesson 4.5 (pp. 169–172) ·

Subtract.

I. What is $15 - 7$?

Step I

 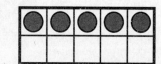

Step 2

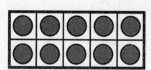 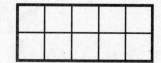

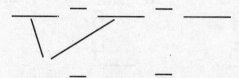

___ − ___ = ___

So, $15 - 7 =$ ___.

School-Home Letter

Dear Family,

My class started Chapter 5 this week. In this chapter, I will learn how addition and subtraction are related. I will also learn how to identify and use related facts.

Love, _____

Vocabulary

related facts Related facts are facts that have the same parts and whole.

$$5 + 7 = 12$$
$$12 - 7 = 5$$

Home Activity

Make a poster with your child like the one below. Each day, write a different related fact. As you progress, leave spaces blank for your child to find missing numbers.

$$\_\_ + 3 = 5 \qquad 5 - 2 = 3$$
$$3 + \_\_ = 5 \qquad 5 - 3 = 2$$

Literature

Look for these books in a library. Read them together to reinforce learning.

Elevator Magic
by Stuart J. Murphy.
HarperCollins, 1997.

Animals on Board
by Stuart J. Murphy.
HarperCollins, 1998.

Carta
para la casa

Querida familia:

Mi clase comenzó el Capítulo 5 esta semana. En este capítulo, aprenderé cómo se relacionan la suma y la resta. También aprenderé cómo identificar las operaciones relacionadas.

Con cariño, _____

Vocabulario

Operaciones relacionadas Las operaciones relacionadas son operaciones que usan los mismos números.

$$5 + 7 = 12$$
$$12 - 7 = 5$$

Actividad para la casa

Haga un cartel con su hijo como el que está abajo. Cada día, escriba una operación relacionada distinta. A medida que avanzan, deje espacios en blanco para que su hijo complete los números que faltan.

$$\underline{} + 3 = 5 \qquad 5 - 2 = 3$$
$$3 + \underline{} = 5 \qquad 5 - 3 = 2$$

Literatura

Busque estos libros en una biblioteca. Léanlos juntos para reforzar el aprendizaje.

Elevator Magic
por Stuart J. Murphy.
Harper Collins, 1997.

Animals on Board
Por Stuart J. Murphy.
Harper Collins, 1998.

Name _____

Problem Solving • Add or Subtract

Make a model to solve.

1. Stan has 12 pennies.

   Some pennies are new.

   4 pennies are old.

   How many pennies are new?

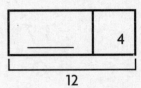

_____ new pennies

2. Liz has 9 toy bears.

   Then she buys some more.

   Now she has 15 toy bears.

   How many toy bears did she buy?

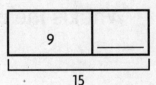

_____ toy bears

3. Eric buys 6 books.

   Now he has 16 books.

   How many books did he have to start?

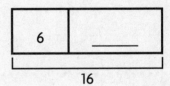

_____ books

4. Cho has 10 rings.

   Some rings are silver.

   4 rings are gold.

   How many rings are silver?

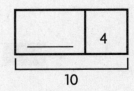

_____ silver rings

Lesson Check

1. Arlo has 17 bean bag animals.
   Some are fuzzy.
   9 bean bag animals are not fuzzy.
   How many animals are fuzzy?

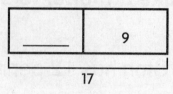

    17   9   8   7

    ○   ○   ○   ○

Spiral Review

2. Count back 3.
   What is the difference? (Lesson 4.1)

$$\underline{\hspace{1cm}} = 11 - 3$$

    9   8   7   6

    ○   ○   ○   ○

3. Which shows a way to make 10? (Lesson 1.7)

    ○

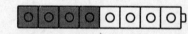

    ○

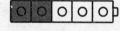

    ○

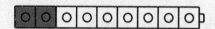

    ○

Algebra • Use Related Facts

Write the missing numbers.

1. Find $16 - 9$.

    $9 + \boxed{} = 16$

    $16 - 9 = \boxed{}$

    $\boxed{16}$ — $\boxed{9}$ — $\boxed{}$

2. Find $12 - 7$.

    $7 + \boxed{} = 12$

    $12 - 7 = \boxed{}$

    $\boxed{12}$ — $\boxed{7}$ — $\boxed{}$

3. Find $15 - 6$.

    $6 + \boxed{} = 15$

    $15 - 6 = \boxed{}$

    $\boxed{15}$ — $\boxed{6}$ — $\boxed{}$

4. Find $18 - 9$.

    $9 + \boxed{} = 18$

    $18 - 9 = \boxed{}$

    $\boxed{18}$ — $\boxed{9}$ — $\boxed{}$

PROBLEM SOLVING

Look at the shapes in the addition sentence.
Draw a shape to show a related subtraction fact.

5.

⬕ + ▽ = ▰   ▰ − _____ = ⬕

Lesson Check

1. Which addition fact helps you solve 12 − 4?

$4 + 6 = 10$
○

$8 + 4 = 12$
○

$7 + 4 = 11$
○

$9 + 3 = 12$
○

Spiral Review

2. Count on to find 9 + 3. (Lesson 3.2)

13
○

12
○

11
○

6
○

3. Which doubles fact matches the picture? (Lesson 3.3)

$\begin{array}{r} 8 \\ + 8 \\ \hline 16 \end{array}$
○

$\begin{array}{r} 7 \\ + 7 \\ \hline 14 \end{array}$
○

$\begin{array}{r} 6 \\ + 6 \\ \hline 12 \end{array}$
○

$\begin{array}{r} 5 \\ + 5 \\ \hline 10 \end{array}$
○

Choose an Operation

Circle add or subtract.
Write a number sentence to solve.

1. Adam has a bag of 11 pretzels.
   He eats 2 of the pretzels.
   How many pretzels are left?

   add   subtract

   _____

   _____ pretzels

2. Greta makes 3 drawings.
   Kate makes 4 more drawings
   than Greta. How many
   drawings does Kate make?

   add   subtract

   _____

   _____ drawings

PROBLEM SOLVING

Choose a way to solve.
Write or draw to explain.

3. Greg has 11 shirts.
   3 have long sleeves.
   The rest have short sleeves.
   How many short-sleeve
   shirts does Greg have?

   _____ short-sleeve shirts

Lesson Check

1. There are 18 children on the bus. Then 9 children get off. Which number sentence shows how to find the number of children left on the bus?

$9 + 8 = 17$ ○ $18 - 9 = 9$ ○

$18 - 0 = 18$ ○ $9 - 9 = 0$ ○

Spiral Review

2. Mike has 13 plants.
   He gives some away.
   He has 4 left.
   How many plants
   does he give away? (Lesson 4.6)

   10 ○   9 ○   7 ○   4 ○

3. What is the sum for $3 + 2 + 8$? (Lesson 3.10)

   13 ○   12 ○   11 ○   4 ○

Algebra • Ways to Make Numbers to 20

Use . Write ways to make the number at the top.

1.   **10**

$$\underline{2} + \underline{7} + \underline{1}$$

$$\underline{5} + \underline{5}$$

$$\underline{10} \underline{} \underline{0}$$

$$\underline{9} \underbigcirc{+} \underline{1}$$

2.   **13**

$$\underline{} + \underline{} + \underline{}$$

$$\underline{} + \underline{}$$

$$\underline{} \bigcirc \underline{}$$

3.   **16**

$$\underline{} + \underline{} + \underline{}$$

$$\underline{} + \underline{}$$

$$\underline{} \bigcirc \underline{}$$

4.   **12**

$$\underline{} + \underline{} + \underline{}$$

$$\underline{} + \underline{}$$

$$\underline{} \bigcirc \underline{}$$

PROBLEM SOLVING

Write numbers to make each line have the same sum.

5.

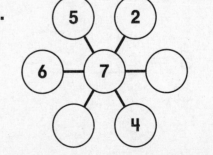

Lesson Check

1. Which way makes 18?

   ○ 4 + 8
   ○ 9 + 0
   ○ 11 − 7
   ○ 8 + 10

2. Which way does **not** make 9?

   ○ 11 + 2
   ○ 11 − 2
   ○ 5 + 4
   ○ 3 + 6

Spiral Review

3. Which is the doubles plus one fact for 7 + 7? (Lesson 3.5)

   ○ 7 + 7 = 14
   ○ 7 + 6 = 13
   ○ 7 + 8 = 15
   ○ 6 + 8 = 14

4. Which is the doubles minus one fact for 4 + 4? (Lesson 3.5)

   ○ 4 + 5 = 9
   ○ 4 + 3 = 7
   ○ 3 + 5 = 8
   ○ 4 + 4 = 8

5. Which subtraction sentence can you solve by using 9 + 5 = 14? (Lesson 4.3)

   14 − 7 = ____
   ○

   14 − 9 = ____
   ○

   9 − 5 = ____
   ○

   14 − 8 = ____
   ○

Algebra • Equal and Not Equal

Which are true? Circle your answers.
Which are false? Cross out your answers.

1. $6 + 4 = 5 + 5$

2. $10 = 6 - 4$

3. $8 + 8 = 16 - 8$

4. $14 = 1 + 4$

5. $8 - 0 = 12 - 4$

6. $17 = 9 + 8$

7. $8 + 3 = 8 - 3$

8. $15 - 6 = 6 + 9$

9. $12 = 5 + 5 + 2$

10. $7 + 6 = 6 + 7$

11. $5 - 4 = 4 + 5$

12. $0 + 9 = 9 - 0$

PROBLEM SOLVING REAL WORLD

13. Which are true? Use a ▭ to color.

$15 = 15$	$12 = 2$	$3 = 8 - 5$
$15 = 1 + 5$	$9 + 2 = 2 + 9$	$9 + 2 = 14$
$1 + 2 + 3 = 3 + 3$	$5 - 3 = 5 + 3$	$13 = 8 + 5$

Lesson Check

1. Which is true?

$4 + 3 = 9 - 2$
○

$4 + 3 = 9 + 2$
○

$4 + 3 = 4 - 3$
○

$4 + 3 = 7 - 3$
○

Spiral Review

2. Which subtraction fact is related to $5 + 6 = 11$? (Lesson 5.3)

$11 - 7 = 4$
○

$11 - 6 = 5$
○

$9 - 3 = 6$
○

$6 - 5 = 1$
○

3. Leah has 4 green toys,
5 pink toys, and 2 blue toys.
How many toys does Leah have? (Lesson 3.12)

6
○

8
○

11
○

12
○

School-Home Letter

Dear Family,

My class started Chapter 6 this week. In this chapter, I will count numbers to 120 and use tens and ones to make numbers.

Love, _____

Vocabulary

ones and **ten** You can group 10 to make 1 ten.

10 ones = 1 ten

hundred 10 tens is the same as 1 hundred.

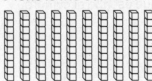

10 tens = 100

Home Activity

Give your child a handful of craft sticks, chenille stems, or straws. Have your child make as many groups of 10 as possible, tying bundles of 10 with a rubber band. Have them place the bundles on a desk or table. Have your child put any leftover ones next to the bundles of 10. Then ask your child to write the number.

Literature

Reading math stories reinforces ideas. Look for these books in a library and read them with your child.

One Is a Snail, Ten Is a Crab by April Pulley Sayre. Candlewick, 2006.

The Counting Family by Jane Manners. Harcourt School Publishers, 2002.

Carta
para la casa

Querida familia:

Mi clase comenzó el Capítulo 6 esta semana. En este capítulo, contaré números hasta el 120 y usaré decenas y unidades para formar números.

Con cariño, _____

Vocabulario

unidades y decenas puedes agrupar unidades para formar decenas

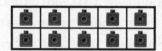

10 unidades = 1 decenas

centena 10 decenas es lo mismo que
1 centena

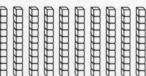

10 decenas = 100

Actividad para la casa

Entréguele a su hijo un puñado de palitos para manualidades, hilos de lana o pajitas. Pídale a su hijo que forme la mayor cantidad de grupos posible, atando paquetes de 10 con una banda elástica. Pídale que coloque los paquetes sobre un escritorio o mesa. Pídale que ponga las unidades sobrantes junto a los paquetes de 10. Luego, pídale a su hijo que escriba el número.

Literatura

Leer cuentos de matemáticas refuerza los conceptos. Busque estos libros en una biblioteca y léalos con su hijo.

One Is a Snail, Ten is a Crab
por April Pulley Sayre.
Candlewick, 2006.

The Counting Family
por Jane Manners.
Harcourt School Publishers, 2002.

Name _____

Count by Ones to 120

Use a Counting Chart. Count forward. Write the numbers.

1. 40, ____, ____, ____, ____, ____, ____, ____, ____

2. 55, ____, ____, ____, ____, ____, ____, ____, ____

3. 37, ____, ____, ____, ____, ____, ____, ____, ____

4. 102, ____, ____, ____, ____, ____, ____, ____, ____

5. 96, ____, ____, ____, ____, ____, ____, ____, ____

PROBLEM SOLVING

Use a Counting Chart. Draw and write numbers to solve.

6. The bag has 111 marbles. Draw more marbles so there are 117 marbles in all. Write the numbers as you count.

Lesson Check

1. Count forward. What number is missing?

110, 111, 112, _____, 114

105	109	113	115
○	○	○	○

Spiral Review

2. There are 6 bees. 2 bees fly away. How many bees are there now? (Lesson 2.1)

2	4	6	8
○	○	○	○

3. There are 8 children. 6 children are boys. The rest are girls. How many children are girls? (Lesson 2.3)

2 girls	3 girls	6 girls	8 girls
○	○	○	○

Lesson Check

1. Count by tens.
   What numbers are missing?

   44, 54, 64, ____, ____, 94

34, 64	44, 45	73, 74	74, 84
○	○	○	○

Spiral Review

2. Which way shows how to
   make a ten to solve 8 + 5? (Lesson 3.9)

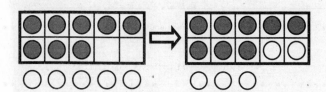

   ○ 8 + 2 + 5

   ○ 6 + 4 + 1

   ○ 9 + 1 + 4

   ○ 8 + 2 + 3

3. Which fact is a related fact? (Lesson 5.2)

   9 + 6 = 15   15 − 6 = 9

   6 + 9 = 15   ☐

   15 − 9 = 6   9 − 6 = 3
   ○   ○

   15 − 6 = 9   7 + 8 = 15
   ○   ○

Count by Tens to 120

Use a Counting Chart.
Count by tens.
Write the numbers.

1. 1, ____, ____, ____, ____, ____, ____, ____, ____, ____

2. 14, ____, ____, ____, ____, ____, ____, ____, ____, ____

3. 7, ____, ____, ____, ____, ____, ____, ____, ____, ____

4. 29, ____, ____, ____, ____, ____, ____, ____, ____, ____

5. 5, ____, ____, ____, ____, ____, ____, ____, ____, ____

6. 12, ____, ____, ____, ____, ____, ____, ____, ____, ____

7. 26, ____, ____, ____, ____, ____, ____, ____, ____, ____

8. 3, ____, ____, ____, ____, ____, ____, ____, ____, ____

9. 8, ____, ____, ____, ____, ____, ____, ____, ____, ____

PROBLEM SOLVING REAL WORLD

Solve.

10. I am after 70.
    I am before 90.
    You say me when you count by tens.
    What number am I?

_____

Name _____

Understand Ten and Ones

Use the model. Write the number three different ways.

1.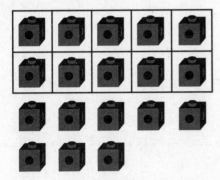

　　　　　　　　_____ ten _____ ones

　　　　　　　　_____ + _____

　　　　　　　　_____

2.

　　　　　　　　_____ ten _____ ones

　　　　　　　　_____ + _____

　　　　　　　　_____

PROBLEM SOLVING REAL WORLD

Draw cubes to show the number.
Write the number different ways.

Rob has 7 ones. Nick has 5 ones. They put all their ones together. What number did they make?

3.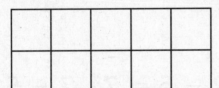

　　　　　　　　_____ ten _____ ones

　　　　　　　　_____ + _____

　　　　　　　　_____

Lesson Check

1. Which shows the same number?

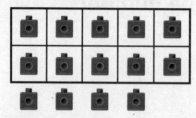

I ten 9 ones   I ten 4 ones   I ten 3 ones   I ten
  ○   ○   ○   ○

Spiral Review

2. What number sentence does this model show? (Lesson 3.7)

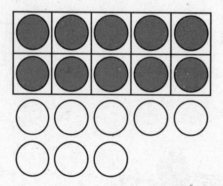

10 − 8 = 2 │ 10 + 8 = 18 │ 5 + 8 = 13 │ 8 + 2 = 10
   ○   ○   ○   ○

3. Which subtraction fact is related to 7 + 5 = 12? (Lesson 5.3)

12 − 6 = 6 │ 14 − 8 = 6 │ 12 − 5 = 7 │ 7 − 5 = 2
   ○   ○   ○   ○

Name _____

Make Ten and Ones

Use . Make groups of ten
and ones. Draw your work.
Write how many.

1.

   14
   fourteen   ____ ten ____ ones

2.

   12
   twelve   ____ ten ____ ones

3.

   15
   fifteen   ____ ten ____ ones

4.

   18
   eighteen   ____ ten ____ ones

5.

   11
   eleven   ____ ten ____ one

PROBLEM SOLVING REAL WORLD

Solve.

6. Tina thinks of a number that has 3 ones and 1 ten.
   What is the number?

   ____

Lesson Check

1. How many tens and ones make this number?

17
seventeen

1 ten 7 ones ○   1 ten 17 ones ○   10 tens 7 ones ○   17 tens ○

Spiral Review

2. What number sentence does this model show? (Lesson 3.7)

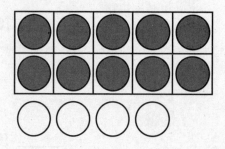

$10 - 4 = 6$ ○ $10 + 5 = 15$ ○ $10 + 4 = 14$ ○ $4 + 1 = 5$ ○

3. Ben has 17 books. He gives some away. He has 8 left. How many books does he give away? (Lesson 4.6)

7 ○   9 ○   10 ○   17 ○

Tens

Use . Make groups of ten.
Write the tens and ones.

1. 90 ones

   ____ tens = ____ ones   ____ tens = ____
                                                                   ninety

2. 50 ones

   ____ tens = ____ ones   ____ tens = ____
                                                                    fifty

3. 40 ones

   ____ tens = ____ ones   ____ tens = ____
                                                                   forty

4. 80 ones

   ____ tens = ____ ones   ____ tens = ____
                                                                   eighty

PROBLEM SOLVING REAL WORLD

Look at the model. Write the number.

5. What number does the model show?

____

Lesson Check

1. What number does the model show?

    20   30   40   50
     ○   ○   ○   ○

2. What number does the model show?

    90   80   70   60
     ○   ○   ○   ○

Spiral Review

3. What is the missing number? (Lesson 5.5)

$$6 + \boxed{} = 13$$

  13   9   8   7
   ○   ○   ○   ○

4. What is the sum for 3 + 3 + 4? (Lesson 3.10)

  6   9   10   12
  ○   ○   ○   ○

Name _____

Tens and Ones to 50

Write the numbers.

1.

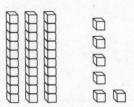

____ tens ____ ones = ____

2.

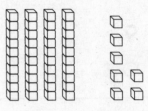

____ tens ____ ones = ____

3.

____ tens ____ ones = ____

4.

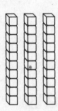

____ tens ____ ones = ____

PROBLEM SOLVING REAL WORLD

Solve. Write the numbers.

5. I have 43 cubes. How many tens and ones can I make?

____ tens ____ ones

Lesson Check

1. Which number does the model show?

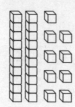

9	20	29	48
○	○	○	○

2. Which number does the model show?

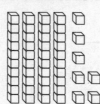

47	40	25	7
○	○	○	○

Spiral Review

3. What is the sum? (Lesson 1.8)

$$\begin{array}{r} 6 \\ + 3 \\ \hline \end{array}$$

12	10	9	6
○	○	○	○

4. What is the difference? (Lesson 2.2)

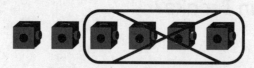

4	3	2	1
○	○	○	○

Tens and Ones to 100

Write the numbers.

1.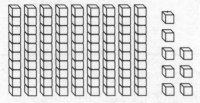

_____ tens _____ ones = _____

....

2.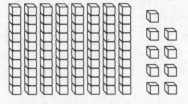

_____ tens _____ ones = _____

....

3.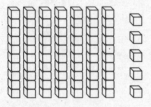

_____ tens _____ ones = _____

....

4.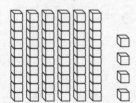

_____ tens _____ ones = _____

PROBLEM SOLVING

Draw a quick picture to show the number.
Write how many tens and ones there are.

5. Inez has 57 shells.

_____ tens _____ ones

Lesson Check

1. What number has 10 tens 0 ones?

     10   20   50   100
      ○   ○   ○   ○

2. What number does
   the model show?

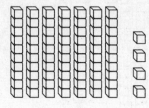

     11   47   74   77
      ○   ○   ○   ○

Spiral Review

3. Which shows the same addends
   in a different order? (Lesson 3.1)

                                    $6 + 5 = 11$

   $6 + 6 = 12$ | $5 + 6 = 11$ | $3 + 4 = 7$ | $5 + 5 = 10$
       ○   ○   ○   ○

4. Count on to solve $2 + 6$. (Lesson 3.2)

      7   8   9   10
      ○   ○   ○   ○

Name _____

Problem Solving • Show Numbers in Different Ways

**Use ▭▭▭▭▭ ▢ to show the number
two different ways. Draw both ways.**

1. 62

Tens	Ones

_____ ◯

Tens	Ones

2. 38

Tens	Ones

_____ ◯

Tens	Ones

3. 47

Tens	Ones

_____ ◯

Tens	Ones

Lesson Check

1. Which is a different way to show the same number?

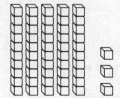

   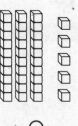   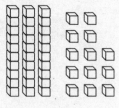

○   ○   ○   ○

2. What number does the model show?

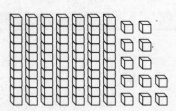

- ○ 77
- ○ 82
- ○ 87
- ○ 92

Spiral Review

3. Which addition sentence can you use to check the subtraction? (Lesson.5.4)

$12 - 4 = \boxed{}$

- ○ $8 + 4 = 12$
- ○ $4 + 4 = 8$
- ○ $3 + 9 = 12$
- ○ $3 + 3 = 6$

4. Which way makes 15? (Lesson 5.8)

$6 + 3$ $7 + 8$ $9 + 2$ $10 - 1$

○   ○   ○   ○

Lesson Check

1. What number does the model show?

14 ○   100 ○   101 ○   104 ○

Spiral Review

2. What is the difference? (Lesson 2.2)

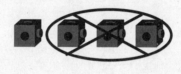

1 ○   3 ○   4 ○   5 ○

3. Ken has 8 pennies. Ron has 3 pennies.
How many fewer pennies does
Ron have than Ken? (Lesson 2.6)

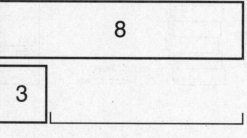

3 ○   4 ○   5 ○   8 ○

Name _____

Model, Read, and Write Numbers from 100 to 110

Use  to show the number.
Write the number.

1. 10 tens and
6 more

_____

2. 10 tens and
1 more

_____

3. 10 tens and
9 more

_____

Write the number.

4.

_____

5.

_____

PROBLEM SOLVING REAL WORLD

6. Solve to find the number of pens.

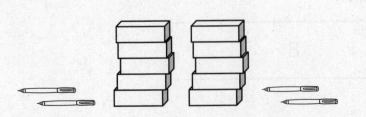

THINK

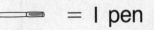 = 1 pen

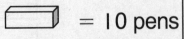 = 10 pens

There are _____ pens.

Name _____

Model, Read, and Write Numbers from 110 to 120

Use  to model the number.
Write the number.

1.

____

2.

____

3.

____

4.

____

5.

____

6.

____

PROBLEM SOLVING REAL WORLD

Choose a way to solve. Draw or write to explain.

7. Dave collects rocks. He makes 12 groups of 10 rocks and has none left over. How many rocks does Dave have?

____ rocks

Lesson Check

1. What number does the model show?

117	115	113	109
○	○	○	○

Spiral Review

2. Which way shows how to make a ten to solve 13 − 7? (Lesson 4.5)

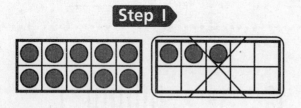

Step 1 **Step 2**

10 − 3	10 − 3 − 4	13 − 3	13 − 3 − 4
○	○	○	○

3. What is the difference? (Lesson 2.9)

$$\begin{array}{r} 9 \\ -\ 4 \\ \hline \end{array}$$

4	5	9	13
○	○	○	○

School-Home Letter

Dear Family,

My class started Chapter 7 this week. In this chapter, I will compare numbers to show greater than or less than. I will also use <, >, and = to compare numbers.

Love, _____

Vocabulary

is greater than > a symbol used to show that a number is greater than another number

$$11 > 10$$
11 is greater than 10

is less than < a symbol used to show that a number is less than another number

$$10 < 11$$
10 is less than 11

Home Activity

Make flash cards for the greater than symbol >, and the less than symbol <. Each day, choose two numbers between 1 and 100. Use the flashcards with your child to compare the numbers.

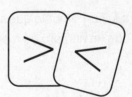

Literature

Look for these books in a library. Use < and > flashcards to compare the groups of objects.

Just Enough Carrots by Stuart J. Murphy. HarperTrophy, 1997.

More, Fewer, Less by Tana Hoban. Greenwillow, 1998.

Carta
para la casa

Querida familia:

Mi clase comenzó el Capítulo 7 esta semana. En este capítulo, compararé números para mostrar los conceptos de mayor que y menor que. También usaré los símbolos <, >, e = para comparar números.

Con cariño, _____

Vocabulario

es mayor que > un símbolo que se usa para mostrar que un número es mayor que otro número

$$11 > 10$$
11 es mayor que 10

es menor que < un símbolo que se usa para mostrar que un número es menor que otro número

$$10 < 11$$
10 es menor que 11

Actividad para la casa

Haga tarjetas nemotécnicas con el símbolo es mayor que > y el símbolo es menor que <. Cada día, elija dos números entre 1 y 100. Use las tarjetas nemotécnicas con su hijo para comparar los números.

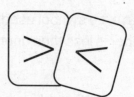

Literatura

Busque estos libros en una biblioteca. Use tarjetas nemotécnicas con < y > para comparar grupos de objetos.

Just Enough Carrots
por Stuart J. Murphy.
Harper Collins, 1997.

More, Fewer, Less
por Tana Hoban.
Greenwillow, 1998.

Algebra • Greater Than

Use  if you need to.

Circle the greater number.	Did tens or ones help you decide?	Write the numbers.
1. 22 42	tens ones	_____ is greater than _____. _____ > _____
2. 46 64	tens ones	_____ is greater than _____. _____ > _____
3. 88 86	tens ones	_____ is greater than _____. _____ > _____
4. 92 29	tens ones	_____ is greater than _____. _____ > _____

PROBLEM SOLVING REAL WORLD

5. Color the blocks that show numbers greater than 47.

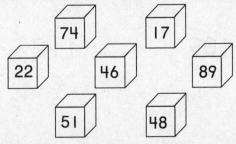

Lesson Check

1. Which number is greater than 65?

37	49	56	66
○	○	○	○

2. Which number is greater than 29?

19	20	28	92
○	○	○	○

Spiral Review

3. What is 5 + 7? (Lesson 5.10)

8	9	11	12
○	○	○	○

4. Count forward. What number is missing? (Lesson 6.1)

110, 111, ___, 113, 114

108	109	112	115
○	○	○	○

Name _____

Algebra • Less Than

Use ⬚⬚⬚⬚⬚⬚ ⬚ if you need to.

Circle the number that is less.	Did tens or ones help you decide?	Write the numbers.
1. 34 36	tens ones	____ is less than ____. ____ < ____
2. 75 57	tens ones	____ is less than ____. ____ < ____
3. 80 89	tens ones	____ is less than ____. ____ < ____
4. 13 31	tens ones	____ is less than ____. ____ < ____

PROBLEM SOLVING

Write a number to solve.

5. Lori makes the number 74. Gabe makes
   a number that is less than 74. What
   could be a number Gabe makes?   ____

Lesson Check

1. Which number is less than 52?

    25 ○   52 ○   64 ○   88 ○

2. Which number is less than 76?

    100 ○   81 ○   77 ○   59 ○

Spiral Review

3. Which number does the model show? (Lesson 6.6)

   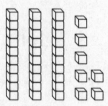

    29 ○   30 ○   37 ○   38 ○

4. Count by tens.
   What numbers are missing?

   (Lesson 6.2)

    8, 18, 28, _____, _____, 58

    19, 29 ○   27, 39 ○   38, 48 ○   46, 57 ○

Lesson Check

1. Which is true?

$22 < 28$ $\bigcirc$ $22 > 28$ $\bigcirc$ $22 = 28$ $\bigcirc$ $28 < 22$ $\bigcirc$

2. Which is true?

$78 > 87$ $\bigcirc$ $78 = 87$ $\bigcirc$ $87 > 78$ $\bigcirc$ $87 < 78$ $\bigcirc$

Spiral Review

3. Which number does the model show? (Lesson 6.7)

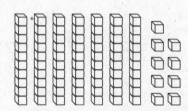

75 $\bigcirc$ 77 $\bigcirc$ 79 $\bigcirc$ 80 $\bigcirc$

4. Which shows the same number?

(Lesson 6.3)

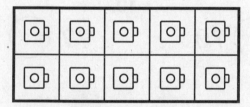

...en 2 ones $\bigcirc$ I ten 3 ones $\bigcirc$ I ten 5 ones $\bigcirc$ I ten 8 ones $\bigcirc$

Algebra • Use Symbols to Compare

Write <, >, or =.
Draw a quick picture if you need to.

1.

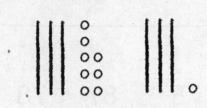

38 ◯ 31

2.

26 ◯ 42

3. 88 ◯ 78

4. 77 ◯ 77

5. 91 ◯ 89

6. 80 ◯ 82

7. 33 ◯ 44

8. 51 ◯ 60

9. 70 ◯ 70

10. 99 ◯ 98

11. 85 ◯

PROBLEM SOLVING REAL WORLD

Write <, >, or = to solve. Circle your answer.

12. Tracey has 26 pennies. Heba has
    29 pennies. Who has a greater
    number of pennies?

    Tracey   Heba

Name _____

Problem Solving •
Compare Numbers

Make a model to solve.

1. Ava has these number cards. She gives away cards with numbers less than 34 and greater than 38. Which number cards does Ava have now?

| 32 | 33 | 35 | 37 | 39 |

Ava has number cards _____.

2. Ron has these number cards. He keeps the cards with numbers greater than 60 and less than 56. Circle the number cards Ron keeps.

| 54 | 57 | 58 | 59 | 61 |

Ron keeps number cards _____.

3. Mia has these number cards. She keeps the cards with numbers less than 85 and greater than 88. Circle the cards Mia keeps.

| 84 | 86 | 87 | 89 | 90 |

Mia keeps number cards _____.

© Houghton Mifflin Harcourt Publishing Company

Lesson Check

1. Juan crosses out the numbers that are less than 45 and greater than 50. Which numbers are left?

| 43 | 44 | 46 | 49 | 52 |

43 and 44   44 and 46   46 and 49   49 and 52
   ○   ○   ○   ○

Spiral Review

2. Count back 1, 2, or 3. What is the difference? (Lesson 4.1)

$$9 - 3 = \underline{\hspace{1cm}}$$

5   6   7   9
○   ○   ○   ○

3. Which completes the related facts? (Lesson 5.2)

$$4 + 7 = 11 \qquad 11 - 4 = 7$$
$$7 + 4 = 11 \qquad \boxed{}$$

$7 + 7 = 14$ $11 - 7 = 4$ $6 + 5 = 11$ $11 - 5 = 6$
○   ○   ○   ○

10 Less, 10 More

Use mental math.
Complete the chart.

	10 Less		10 More
1.	___	48	___
2.	___	25	___
3.	___	73	___
4.	___	89	___
5.	8	___	
6.	___	___	47

PROBLEM SOLVING

Choose a way to solve. Draw or write to show your work.

7. Jim has 16 pennies. Doug has 10 fewer pennies than Jim. How many pennies does Doug have?

_____ pennies

Lesson Check

1. Which number is 10 less than 67?

77   68   66   57

○   ○   ○   ○

2. Which number is 10 more than 39?

49   40   38   29

○   ○   ○   ○

Spiral Review

3. How many tens and ones make this number? (Lesson 6.4)

18
eighteen

8 ones   1 ten 7 ones   1 ten 8 ones   2 tens

○   ○   ○   ○

4. What number does the model show? (Lesson 6.5)

80   70   60   50

○   ○   ○   ○

Chapter 7 Extra Practice

Lesson 7.1 (pp. 289–292) ·

Use ▱▱▱▱▱▱ ▱ if you need to.

Circle the greater number.	Did tens or ones help you decide?	Write the numbers.
1. 46 34	tens ones	___ is greater than ___. ___ > ___
2. 77 79	tens ones	___ is greater than ___. ___ > ___

Lesson 7.2 (pp. 293 – 296) ·

Use ▱▱▱▱▱▱ ▱ if you need to.

Circle the number that is less.	Did tens or ones help you decide?	Write the numbers.
1. 23 29	tens ones	___ is less than ___. ___ < ___
2. 64 95	tens ones	___ is less than ___. ___ < ___

Lesson 7.3 (pp. 297–299)

Write <, >, or =.

Draw a quick picture if you need to.

1. 24 ◯ 42

2. 32 ◯ 22

3. 76 ◯ 76

4. 91 ◯ 81

5. 61 ◯ 63

6. 58 ◯ 58

Lesson 7.5 (pp. 305–308)

Use mental math.

Complete the chart.

	10 **Less**		10 **More**
1.	_____	26	_____
2.	_____	75	_____
3.	_____	44	_____

School-Home Letter

Dear Family,

My class started Chapter 8 this week. In this chapter, I will learn how to add and subtract two-digit numbers.

Love, _____

Vocabulary

ones and tens You can group ones to make tens.

20 ones = 2 tens

Home Activity

Using a jar and pennies, work with your child to add and subtract two-digit numbers. Start with 11 pennies in the jar. Have your child add 13 pennies. Ask your child to explain a way to find the sum of 11 and 13. Then count with your child to find how many pennies in all. Repeat with different quantities each day. Work with your child to subtract numbers as well.

Literature

Reading math stories reinforces ideas. Look for these books in a library and read them with your child.

One Is a Snail, Ten Is a Crab by April Pulley Sayre. Candlewick, 2006.

Safari Park by Stuart J. Murphy. Steck-Vaughn, 2002.

Carta
para la casa

Querida familia:

Mi clase comenzó el Capítulo 8 esta semana. En este capítulo, aprenderé a sumar y restar números de dos dígitos.

Con cariño, _____

Vocabulario

unidades y decenas Puedes agrupar unidades para formar decenas.

20 unidades = 2 decenas

Actividad para la casa

Con un tarro y centavos, trabaje con su hijo para sumar y restar números de dos dígitos. Comience con 11 monedas de 1¢ en el tarro. Pídale que sume 13 monedas de 1¢. Pídale que explique cómo encontrar la suma de 11 y 13. Luego, cuente con su hijo para saber cuántos centavos hay en total. Repita con diferentes cantidades cada día. Trabaje con él para restar números también.

Literatura

Leer cuentos de matemáticas refuerza las ideas. Busque estos libros en la biblioteca y léalos con su hijo.

One Is a Snail, Ten Is a Crab
por April Pulley Sayre. Candlewick, 2006.

Safari Park
por Stuart J. Murphy. Steck-Vaughn, 2002.

Lesson Check

1. What is the sum?

$$8 + 5 = \underline{}$$

3   12   13   16

○   ○   ○   ○

2. What is the difference?

$$11 - 4 = \underline{}$$

6   7   9   15

○   ○   ○   ○

Spiral Review

3. Which number is greater than 40? (Lesson 7.1)

34   40   42   46

○   ○   ○   ○

Which number is less than 84? (Lesson 7.2)

94   93   85   69

○   ○   ○   ○

Add and Subtract within 20

Add or subtract.

1.	2.	3.	4.	5.	6.
6 +0	11 − 2	4 +5	9 +8	4 +10	14 − 9

7.	8.	9.	10.	11.	12.
7 +4	8 −5	12 − 3	6 +7	18 − 9	15 − 6

13.	14.	15.	16.	17.	18.
6 +5	12 − 6	10 −10	13 − 7	2 +7	6 +4

PROBLEM SOLVING REAL WORLD

Solve. Draw or write to explain.

19. Jesse has 4 shells. He finds some more. Now he has 12 shells. How many more shells did Jesse find?

_____ m

Subtract Tens

Draw to show tens. Write the difference. Write how many tens.

1. $40 - 10 =$ _____

_____ tens

2. $80 - 40 =$ _____

_____ tens

3. $50 - 30 =$ _____

_____ tens

4. $60 - 30 =$ _____

_____ tens

PROBLEM SOLVING  REAL WORLD

Draw tens to solve.

5. Mario has 70 baseball cards.
He gives 30 to Lisa.
How many baseball cards
does Mario have left?

_____ baseball cards

Lesson Check

1. What is the difference?

$$60 - 20 = \underline{\hphantom{00}}$$

80 ○   50 ○   40 ○   20 ○

2. What is the difference?

$$70 - 30 = \underline{\hphantom{00}}$$

30 ○   40 ○   50 ○   70 ○

Spiral Review

3. What is the sum? (Lesson 3.8)

$$\begin{array}{r} 9 \\ + 4 \\ \hline \end{array}$$

4 ○   5 ○   6 ○   13 ○

4. Bo crosses out the numbers that are less than 33 and greater than 38. What numbers are left? (Lesson 7.4)

| 30 | 32 | 36 | 37 | 39 |

30 and 32 ○   32 and 36 ○   36 and 37 ○   37 and 39 ○

Name _____

Use a Hundred Chart to Add

Use the hundred chart to add.
Count on by ones or tens.

1. $47 + 2 =$ _____

....

2. $26 + 50 =$ _____

....

3. $22 + 5 =$ _____

....

4. $40 + 41 =$ _____

....

5. $4 + 85 =$ _____

1	2	3	4	5	6	7	8	9	10
11	12	13	14	15	16	17	18	19	20
21	22	23	24	25	26	27	28	29	30
31	32	33	34	35	36	37	38	39	40
41	42	43	44	45	46	47	48	49	50
51	52	53	54	55	56	57	58	59	60
61	62	63	64	65	66	67	68	69	70
71	72	73	74	75	76	77	78	79	80
81	82	83	84	85	86	87	88	89	90
91	92	93	94	95	96	97	98	99	100

PROBLEM SOLVING  REAL WORLD

Choose a way to solve. Draw or write to show your work.

6. 17 children are on the bus.
Then 20 more children get on
the bus. How many children
are on the bus now?

_____ children

© Houghton Mifflin Harcourt Publishing Company

Lesson Check

1. What is 42 + 50?

47 ○   82 ○

92 ○   97 ○

1	2	3	4	5	6	7	8	9	10
11	12	13	14	15	16	17	18	19	20
21	22	23	24	25	26	27	28	29	30
31	32	33	34	35	36	37	38	39	40
41	42	43	44	45	46	47	48	49	50
51	52	53	54	55	56	57	58	59	60
61	62	63	64	65	66	67	68	69	70
71	72	73	74	75	76	77	78	79	80
81	82	83	84	85	86	87	88	89	90
91	92	93	94	95	96	97	98	99	100

2. What is 11 + 8?

19 ○   81 ○

91 ○   99 ○

Spiral Review

3. What number is ten less than 52? (Lesson 7.5)

50 ○   48 ○   42 ○   41 ○

4. Which addition fact helps
you solve 16 − 9? (Lesson 5.6)

6 + 9 = 15 ○   9 + 7 = 16 ○

8 + 8 = 16 ○   9 + 9 = 18 ○

Make Ten to Add

Use ⬜⬜⬜⬜⬜⬜⬜⬜⬜⬜ ⬜. Draw to show how
you make a ten. Find the sum.

1. $26 + 5 =$ _____

2. $68 + 4 =$ _____

3. $35 + 8 =$ _____

PROBLEM SOLVING REAL WORLD

Choose a way to solve. Draw or
write to show your work.

4. Debbie has 27 markers. Sal
has 9 markers. How many
markers do they have?

_____ markers

Lesson Check

1. What is 47 + 6?

53   54   56   57
○   ○   ○   ○

2. What is 84 + 8?

91   92   94   96
○   ○   ○   ○

Spiral Review

3. What number does the model show? (Lesson 6.10)

100   104   114   140
○   ○   ○   ○

4. Which makes the sentence true? (Lesson 5.9)

$$5 + 4 = 10 - \underline{\hspace{2em}}$$

1   3   4   5
○   ○   ○   ○

Name _____

Use Place Value to Add

Draw a quick picture. Use tens and ones to add.

1.

$$\begin{array}{r} 31 \\ + 26 \\ \hline \end{array}$$

Tens	Ones

3 tens + 1 one
2 tens + 6 ones
_____
___ tens + ___ ones

___ + ___ = ___

$$\begin{array}{r} 31 \\ + 26 \\ \hline \end{array}$$

2.

$$\begin{array}{r} 54 \\ + 34 \\ \hline \end{array}$$

Tens	Ones

5 tens + 4 ones
3 tens + 4 ones
_____
___ tens + ___ ones

___ + ___ = ___

$$\begin{array}{r} 54 \\ + 34 \\ \hline \end{array}$$

PROBLEM SOLVING

3. Write two addition sentences you can use to find the sum. Then solve.

Addend   **Addend**

___ + ___ = ___

___ + ___ = ___

Lesson Check

1. What is the sum?

$$\begin{array}{r} 42 \\ + 31 \\ \hline \end{array}$$

11 ○   13 ○   73 ○   74 ○

2. What is the sum?

$$\begin{array}{r} 23 \\ + 12 \\ \hline \end{array}$$

11 ○   25 ○   32 ○   35 ○

Spiral Review

3. What number is the same as 2 tens 8 ones? (Lesson 6.6)

82 ○   28 ○   10 ○   8 ○

4. What is the sum? (Lesson 3.3)

$$\begin{array}{r} 5 \\ + 5 \\ \hline \end{array}$$

9 ○   10 ○   11 ○   12 ○

Name _____

Problem Solving • Addition Word Problems

Draw and write to solve. Explain your reasoning.

1. Dale saved 19 pennies. Then he found 5 more pennies. How many pennies does Dale have now?

   _____

   _____

   _____

   _____ pennies

2. Jean has 10 fish. She gets 4 more fish. How many fish does she have now?

   _____

   _____

   _____

   _____ fish

3. Courtney buys 2 bags of apples. Each bag has 20 apples. How many apples does she buy?

   _____

   _____

   _____

   _____ apples

4. John bakes 18 blueberry muffins and 12 banana muffins for the bake sale. How many muffins does he bake?

   _____

   _____

   _____

   _____ muffins

© Houghton Mifflin Harcourt Publishing Company

Lesson Check

1. Amy has 9 books about dogs.
   She has 13 books about cats.
   How many books does she
   have about dogs and cats?

   20   22   23   25
   ○   ○   ○   ○

Spiral Review

2. What is the sum for 4 + 2 + 4? (Lesson 3.10)

   12   11   10   9
   ○   ○   ○   ○

3. Which shows a way
   to make a ten to subtract? (Lesson 4.4)

   $$14 - 8 = ?$$

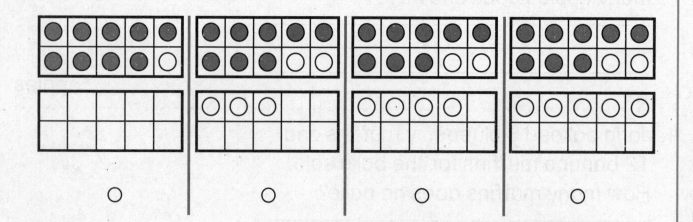

   ○   ○   ○   ○

Name _____

Practice Addition and Subtraction

Add or subtract.

1.	20 + 20	2.	90 − 30	3.	52 + 4	4.	62 + 21	5.	39 − 10
6.	8 + 2	7.	47 + 34	8.	4 − 0	9.	49 − 6	10.	64 + 30
11.	63 + 11	12.	37 − 6	13.	85 + 13	14.	48 + 11	15.	76 − 15

PROBLEM SOLVING

Solve. Write or draw to explain.

16. Andrew read 17 pages of his book before dinner. He read 9 more pages after dinner. How many pages did he read?

_____ pages

Lesson Check

1. Which is the sum of 20 + 18?

   40   38   28   2
   ○   ○   ○   ○

2. Which is the difference of 90 − 50?

   60   50   40   30
   ○   ○   ○   ○

Spiral Review

3. What number sentence does this model show? **(Lesson 3.7)**

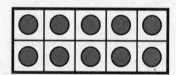

○ ○ ○

$10 - 3 = 7$ $10 + 3 = 13$
  ○   ○

$5 + 3 = 8$ $3 + 1 = 4$
  ○   ○

4. Mo had some toys. He gave 6 away. Now he has 6 toys. How many toys did Mo start with? **(Lesson 4.6)**

   0   6   10   12
   ○   ○   ○   ○

Chapter 8 Extra Practice

Name _____

Lesson 8.1 (pp. 317 – 320) ·
Add or subtract.

1. $10 + 3 =$ _____

2. $16 - 9 =$ _____

Lessons 8.2 - 8.3 (pp. 321 – 327) ·
Draw to show tens. Write the sum or difference. Write how many tens.

1. $30 + 60 =$ _____

2. $70 - 20 =$ _____

_____ tens

_____ tens

Lesson 8.4 (pp. 329 – 332) · · · · · · · · · · · · · · · · · · ·
Use the hundred chart to add.
Count on by ones or tens.

1	2	3	4	5	6	7	8	9	10
11	12	13	14	15	16	17	18	19	20
21	22	23	24	25	26	27	28	29	30
31	32	33	34	35	36	37	38	39	40
41	42	43	44	45	46	47	48	49	50
51	52	53	54	55	56	57	58	59	60
61	62	63	64	65	66	67	68	69	70
71	72	73	74	75	76	77	78	79	80
81	82	83	84	85	86	87	88	89	90
91	92	93	94	95	96	97	98	99	100

1. $81 + 6 =$ _____

· ·

2. $75 + 20 =$ _____

· ·

3. $30 + 42 =$ _____

Lesson 8.5 (pp. 333 – 336)

Use ▭ ▱ and your MathBoard.
Add the ones or tens.
Write the sum.

1. $35 + 30 =$ _____

2. $3 + 71 =$ _____

3. $44 + 5 =$ _____

4. $20 + 11 =$ _____

Lessons 8.6 - 8.7 (pp. 337 – 344)

Write the sum.

1. $56 + 8 =$ _____

2. $5 + 27 =$ _____

3. $13 + 7 =$ _____

4. $33 + 9 =$ _____

5. $6 + 64 =$ _____

6.
$$\begin{array}{r} 61 \\ + 29 \\ \hline \end{array}$$

7.
$$\begin{array}{r} 73 \\ + 18 \\ \hline \end{array}$$

Lesson 8.9 (pp. 349 – 352)

Add or subtract.

1. $16 - 8 =$ _____

2. $35 + 53 =$ _____

3.
$$\begin{array}{r} 48 \\ - 5 \\ \hline \end{array}$$

4.
$$\begin{array}{r} 10 \\ + 80 \\ \hline \end{array}$$

5.
$$\begin{array}{r} 79 \\ - 9 \\ \hline \end{array}$$

6.
$$\begin{array}{r} 3 \\ - 3 \\ \hline \end{array}$$

School-Home Letter

Dear Family,

My class started Chapter 9 this week. In this chapter, I will learn about measurement. I will use length to compare, order, and measure objects. I will also use time to tell time to the hour and half hour.

Love, _____

Vocabulary

hour

half hour

Home Activity

Cut strips of paper in varying lengths and place them in random order on a table. Have children put the strips of paper in order from longest to shortest.

Literature

Look for these books in a library.

How Big Is a Foot?
Rolf Myller.
Dell Yearling, 1991.

Super Sand Castle Saturday
Stuart J. Murphy.
HarperTrophy, 1999.

Carta
para la casa

Querida familia:

Mi clase comenzó el Capítulo 9 esta semana. En este capítulo, aprenderé sobre medidas. Usaré la longitud para comparar, ordenar y medir objetos. También usaré el tiempo para decir la hora y la media hora.

Con cariño, _____

Vocabulario

hora

media hora

Actividad para la casa

Corte tiras de papel que tengan una longitud variada y colóquelas sobre una mesa en orden aleatorio. Pídales a los niños que pongan las tiras de papel en orden, de la más larga a la más corta.

Literatura

Busque estos libros en una biblioteca.

How Big Is a Foot?
por Rolf Myller.
Dell Yearling, 1991.

Sábado de super castillos
por Stuart J. Murphy.
HarperTrophy, 1998.

Order Length

Draw three pencils in order from
shortest to longest.

1. shortest |

2. |

3. longest |

Draw three markers in order from
longest to shortest.

4. longest |

5. |

6. shortest |

PROBLEM SOLVING  REAL WORLD

Solve.

7. Fred has the shortest
toothbrush in the bathroom.
Circle Fred's toothbrush.

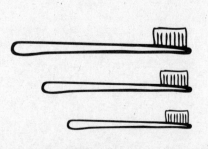

Lesson Check

1. Which line is the longest?

○

○

○

○

2. Which paintbrush is the shortest?

○

○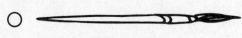

○

○ ▬▬▬▬

Spiral Review

3. Which is a different way to show the same number? **(Lesson 6.8)**

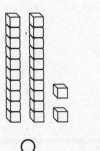

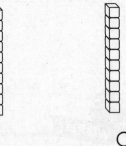

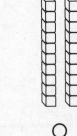

○   ○   ○   ○

Name _____

Indirect Measurement

Read the clues. Write shorter or longer to complete the sentence. Then draw to prove your answer.

1. Clue 1: A yarn is longer than a ribbon.

   Clue 2: The ribbon is longer than a crayon.

   So, the yarn is _____ than the crayon.

yarn

ribbon

crayon

PROBLEM SOLVING REAL WORLD

Solve. Draw or write to explain.

2. Megan's pencil is shorter than Tasha's pencil.

   Tasha's pencil is shorter than Kim's pencil.

   Is Megan's pencil shorter or longer than Kim's pencil?

   _____

Lesson Check

1. A black line is longer than the gray line. The gray line is longer than a white line. Which is correct?

○

○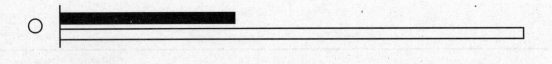

○

○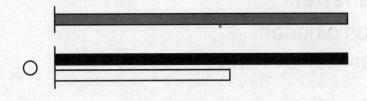

....

Spiral Review

2. What is the sum? (Lesson 8.4)

$$42 + 20 = \underline{\hspace{1cm}}$$

62   44   40   22
○   ○   ○   ○

Name _____

Use Nonstandard Units to Measure Length

Use real objects. Use  to measure.

1.

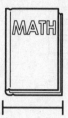

about _____

2.

about _____

3.

about _____

4.

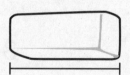

about _____

PROBLEM SOLVING REAL WORLD

Solve.

5. Don measures his desk with .
About how long is his desk?

about _____

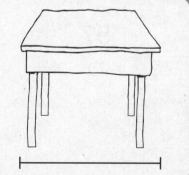

© Houghton Mifflin Harcourt Publishing Company

Lesson Check

1. Use ■. Kevin measures the ribbon with ■.
   About how long is the ribbon?

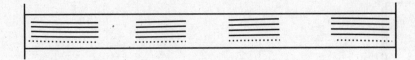

- ○ about 1 ■ long
- ○ about 3 ■ long
- ○ about 4 ■ long
- ○ about 6 ■ long

Spiral Review

2. I have 27 red flowers and
   19 white flowers. How many
   flowers do I have? (Lesson 8.8)

47	46	37	36
○	○	○	○

3. What number is less than 51? (Lesson 7.2)

57	55	52	50
○	○	○	○

Make a Nonstandard Measuring Tool

Use the measuring tool you made.
Measure real objects.

1.

about _____

2.

about _____

3.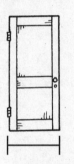

about _____

4.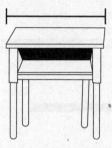

about _____

5.

about _____

6.

about _____

7.

about _____

8.

about _____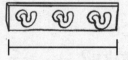

Lesson Check

1. Use the ⊂⊃ below. Which string is about 4 ⊂⊃ long?

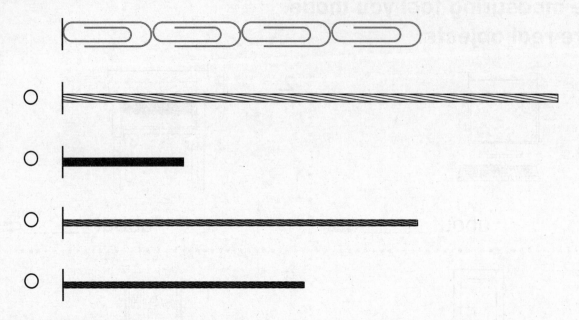

Spiral Review

2. Ty crosses out the numbers
that are greater than 38 and less
than 34. What numbers are left? (Lesson 7.4)

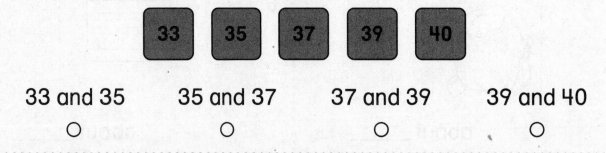

33 and 35   35 and 37   37 and 39   39 and 40
   ○   ○   ○   ○

3. There are 12 books. 4 books are large.
The rest are small. Which number sentence
shows how to find the number of small books? (Lesson 5.7)

○ 11 + 1 = 12   ○ 4 + 6 = 10
○ 12 − 3 = 9   ○ 12 − 4 = 8

Name _____

Problem Solving • Measure and Compare

The blue string is about 3 ☐ long.
The green string is 2 ☐ longer than the blue
string. The red string is 1 ☐ shorter than the
blue string. Measure and draw the strings in
order from **longest** to **shortest**.

1. |

                                                about ____ ☐

2. |

                                                about ____ ☐

3. |

                                                about ____ ☐

PROBLEM SOLVING  REAL WORLD

4. Sandy has a ribbon about 4 ☐ long.
   She cut a new ribbon 2 ☐ longer.
   Measure and draw the two ribbons.

   |
   |

                 The new ribbon is about ____ ☐ long.

Lesson Check

1. Mia measures a stapler with her paper clip ruler. About how long is the stapler?

about 2 ⬭   about 5 ⬭   about 7 ⬭   about 20 ⬭

○   ○   ○   ○

Spiral Review

2. What is the missing number? (Lesson 8.1)

$$4 + 9 = \underline{\hspace{1cm}}$$

13   9   8   4

○   ○   ○   ○

3. Count by tens. What numbers are missing? (Lesson 6.2)

$$17, 27, \underline{\hspace{1cm}}, \underline{\hspace{1cm}}, 57, 67$$

77, 87   28, 29   37, 38   37, 47

○   ○   ○   ○

Time to the Hour

**Look at where the hour hand points.
Write the time.**

1.

_____

2.

_____

3.

_____

4.

_____

5.

_____

6.

_____

PROBLEM SOLVING REAL WORLD

Solve.

7. Which time is **not** the same? Circle it.

 7:00   7 o'clock

Lesson Check

1. Look at the hour hand. What is the time?

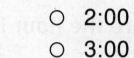

- ○ 2:00
- ○ 3:00
- ○ 4 o'clock
- ○ 5 o'clock

2. Look at the hour hand. What is the time?

- ○ 11:00
- ○ 10:00
- ○ 9 o'clock
- ○ 8 o'clock

Spiral Review

3. What is the sum? (Lesson 8.2)

$40 + 30 =$ ___

10	55	70	84
○	○	○	○

4. What is the sum? (Lesson 8.5)

$53 + 30 =$ ___

83	80	62	23
○	○	○	○

Time to the Half Hour

Look at where the hour hand points. Write the time.

1.

_____

- - - - - - - -

2.

_____

- - - - - - - -

3.

_____

- - - - - - - -

4.

_____

- - - - - - - -

5.

_____

- - - - - - - -

6.

_____

- - - - - - - -

PROBLEM SOLVING REAL WORLD

Solve.

7. Greg rides his bike at half past 4:00. He eats dinner at half past 6:00. He reads a book at half past 8:00.

Look at the clock.
Write what Greg does.

Greg _____.

Lesson Check

1. Look at the hour hand. What is the time?

○ 5:00
○ half past 5:00
○ 6:00
○ half past 6:00

2. Look at the hour hand. What is the time?

○ 10:00
○ half past 10:00
○ half past 9:00
○ 9:00

Spiral Review

3. What number does the model show? (Lesson 6.9)

102   103   107   113
○   ○   ○   ○

4. How many tens and ones make this number? (Lesson 6.4)

14
fourteen

2 tens 4 ones   1 ten 5 ones   1 ten 4 ones   1 ten 2 ones
○   ○   ○   ○

Tell Time to the Hour and Half Hour

Write the time.

1.

2.

3.

4.

5.

6.

PROBLEM SOLVING REAL WORLD

Solve.

7. Lulu walks her dog at 7 o'clock. Bill walks his dog 30 minutes later. Draw to show what time Bill walks his dog.

© Houghton Mifflin Harcourt Publishing Company

Lesson Check

1. What time is it?

- ○ 6:30
- ○ 7:00
- ○ 7:30
- ○ 8:30

2. What time is it?

- ○ 12:00
- ○ 2:00
- ○ 2:30
- ○ 3:30

Spiral Review

3. What is the sum? (Lesson 8.4)

$$48 + 20 = \underline{\hspace{1cm}}$$

- ○ 69
- ○ 68
- ○ 60
- ○ 28

4. How many tens and ones are in the sum? (Lesson 8.7)

$$\begin{array}{r} 67 \\ + 25 \\ \hline \end{array}$$

- ○ 9 tens 2 ones
- ○ 8 tens 7 ones
- ○ 8 tens 2 ones
- ○ 4 tens 2 ones

Name _____

Practice Time to the Hour and Half Hour

**Use the hour hand to write the time.
Draw the minute hand.**

1.

2.

3.

4.

5.

6.

PROBLEM SOLVING REAL WORLD

Solve.

7. Billy played outside for a half hour.
Write how many minutes Billy
played outside.

_____ minutes

© Houghton Mifflin Harcourt Publishing Company

Lesson Check

1. Which clock shows 11:00?

○ ○ ○ ○

Spiral Review

2. What is the difference? (Lesson 8.3)

$$80 - 30 = \underline{\hspace{1cm}}$$

65	55	50	30
○	○	○	○

3. Use . Amy measures the eraser with ■.
   About how long is the eraser? (Lesson 9.3)

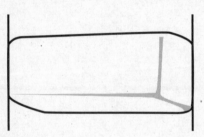

○ about 1 ■ long

○ about 2 ■ long

○ about 3 ■ long

○ about 4 ■ long

Chapter 9 Extra Practice

Lesson 9.1 (pp. 369–372) ·

Draw three paint brushes in order
from **shortest** to **longest**.

I.   **shortest**

   longest

Lesson 9.2 (pp. 373–376) ·

Read the clues. Write **shorter** or **longer**
to complete the sentence. Then draw to
prove your answer.

1. Clue 1: A gray line is longer than a white line.
   Clue 2: A white line is longer than a black line.

So, the gray line is _____ than the black line.

black	
white	
gray	

Lesson 9.3 (pp. 377–380) ·

Use real objects. Use ■ to measure.

I.

about _____

Lesson 9.4 (pp. 381–384) . . . .

Use the measuring tool you made.
Measure real objects.

1.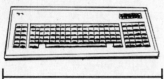

about _____

Lessons 9.6 – 9.7 (pp. 389–396) . . . .

Look at where the hour hand points. Write the time.

1.

2.

3.

_____
– – – – – – – – – – –
_____

Lessons 9.8 – 9.9 (pp. 397–404) . . . .

Write the time.

1.

2.

3.

School-Home Letter

Dear Family,

My class started Chapter 10 this week. In this chapter, I will show data with tally charts and graphs. I will also ask and answer questions about the charts and graphs.

Love, _____

Vocabulary

bar graph a graph that uses bars to show information

picture graph a graph that uses pictures to show information

tally chart a chart that uses tally marks to record information

tally mark a line that stands for one person or thing

Home Activity

Help your child keep track of the weather on a calendar for a week or longer. Then help your child use the data to make a picture graph. Use the graph to compare the number of days that were sunny, cloudy, and rainy.

Weather This Week								
sunny	○	○	○	○				
cloudy	○	○	○					
raniy	○							

Each ○ stands for 1 day.

Literature

Look for these books in a library. These books will reinforce your child's understanding of data and graphs.

The Great Graph Contest by Loreen Leedy. Holiday House, 2006.

Graphing Favorite Things by Jennifer Marrewa. Weekly Reader® Books, 2008.

Carta
para la casa

Querida familia:

Mi clase comenzó el Capítulo 10 esta semana. En este capítulo, mostraré datos con tablas de conteo y gráficas. También haré y responderé preguntas sobre tablas y gráficas.

Con cariño, _____

Vocabulario

gráfica de barras una gráfica que utiliza barras para mostrar información

pictografía una gráfica que utiliza dibujos para mostrar información

tabla de conteo una tabla que utiliza marcas para registrar información

marca de conteo una línea que representa una persona o una cosa

Actividad para la casa

Ayude a su hijo para que siga el clima usando un calendario durante una semana o más. Luego ayude a su hijo para que use los datos para hacer una pictografía. Usen la gráfica para comparar el número de días soleados, nublados y lluviosos.

Weather This Week							
sunny	○	○	○	○			
cloudy	○	○	○				
rainy	○						

Cada ○ representa I diá.

Literatura

Busque estos libros en una biblioteca. Estos libros reforzarán el aprendizaje de su hijo sobre datos y gráficas.

The Great Graph Contest por Loreen Leedy. Holiday House, 2006.

Graphing Favorite Things por Jennifer Marrewa. Weekly Reader® Books, 2008. Albert Whitman and Company, 1993.

Name _____

Read Picture Graphs

Our Favorite Outdoor Activity									
biking	♀	♀	♀	♀	♀	♀	♀	♀	
skating	♀	♀							
running	♀	♀	♀	♀					

Each ♀ stands for 1 child.

Use the picture graph to answer the question.

1. How many children chose 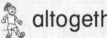?

____ children

2. How many children chose  and 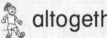 altogether?

____ children

3. Which activity did the most children choose? Circle.

PROBLEM SOLVING REAL WORLD

Write a number sentence to solve the problem.
Use the picture graph at the top of the page.

4. How many more children chose  than 🏃 ?

____ more children

__ __ ◯ __ __ ◯ __ __

Lesson Check

Use the picture graph to answer the question.

Do you do chores at home?								
yes	☺	☺	☺	☺	☺	☺	☺	☺
no	☺	☺	☺	☺	☺	☺		

Each ☺ stands for 1 child.

1. How many children do chores at home?

    2 children   6 children   8 children   14 children
        ○   ○   ○   ○

2. How many more children answered yes than no?

    2 more   6 more   8 more   16 more
       ○   ○   ○   ○

Spiral Review

3. What number is ten less than 82? (Lesson 7.5)

    92   83   81   72
    ○   ○   ○   ○

4. Count forward. What number is missing? (Lesson 6.1)

    110, 111, 112, _____, 114

    100   113   114   115
    ○   ○   ○   ○

Name _____

Make Picture Graphs

Which dinosaur do the most children
like best? Ask 10 friends.
Draw 1 circle for each child's answer.

Our Favorite Dinosaur										
🦖 Tyrannosaurus										
🐃 Triceratops										
🦕 Apatosaurus										

Each ○ stands for 1 child.

Use the picture graph to answer the question.

1. How many children
   chose ?

   _____ children

2. How many children chose
    and altogether?

   _____ children

3. Which dinosaur did the
   fewest children choose?
   Circle.

4. Which dinosaur did the
   most children choose?
   Circle.

PROBLEM SOLVING REAL WORLD

5. Write your own question about the graph.

   _____

   _____

   _____

   _____

Lesson Check

Use the picture graph to answer the question.

Which Hand Do You Use to Eat?								
left	○	○	○					
right	○	○	○	○	○	○	○	○

Each ○ stands for 1 child.

1. How many children use their right hand?

     3   6   8   11
     ○   ○   ○   ○

2. How many more children use their right hand than their left?

     3 more   5 more   6 more   8 more
     ○   ○   ○   ○

Spiral Review

3. What is the sum? (Lesson 8.1)

$$6 + 3 = \underline{\hspace{2cm}}$$

     3   6   8   9
     ○   ○   ○   ○

4. What is the difference of $60 - 20$? (Lesson 8.3)

     8   40   62   80
     ○   ○   ○   ○

Read Bar Graphs

Use the bar graph to answer the question.

1. How many children chose ◯?

_____ children

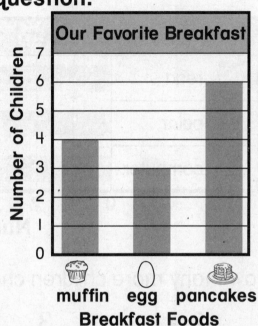

Our Favorite Breakfast

(Number of Children: muffin = 4, egg = 3, pancakes = 6)

Breakfast Foods

2. How many children chose 🧁?

_____ children

3. How many children chose 🧁 or ◯?

_____ children

4. How many more children chose 🥞 than ◯?

_____ more children

PROBLEM SOLVING REAL WORLD

Use the bar graph to answer the question.

5. Claudette uses an ☂.
   Add her to the graph. Now how many more children use an ☂ than a 🎩?

_____ more children

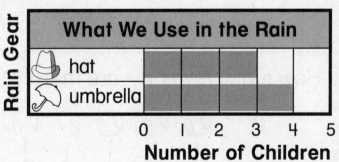

What We Use in the Rain

Rain Gear

hat — 3
umbrella — 3

Number of Children (0 1 2 3 4 5)

Lesson Check

Use the bar graph to answer the question.

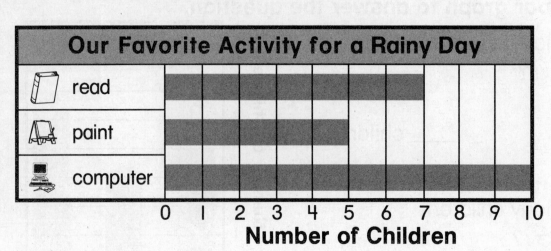

Kinds of Activities

Our Favorite Activity for a Rainy Day

read

paint

computer

0 1 2 3 4 5 6 7 8 9 10

Number of Children

1. How many more children chose 🖥 than 🎨?

   2   3   5   15
   ○   ○   ○   ○

Spiral Review

2. Which subtraction sentence can you solve by using $9 + 7 = 16$? (Lesson 4.3)

   ○ $9 - 7 = \underline{\quad}$

   ○ $9 - 6 = \underline{\quad}$

   ○ $10 - 7 = \underline{\quad}$

   ○ $16 - 7 = \underline{\quad}$

3. How many fewer ⚾ are there ? (Lesson 2.5)

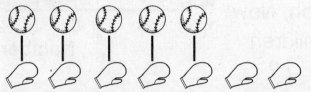

$7 - 5 = \underline{\quad}$

1 fewer ⚾   2 fewer ⚾   3 fewer ⚾   12 fewer ⚾
   ○   ○   ○   ○

Make Bar Graphs

Which is your favorite meal?

1. Ask 10 friends which meal they like best.
   Make a bar graph.

Our Favorite Meal										
breakfast										
lunch										
dinner										

Meal

0   1   2   3   4   5   6   7   8   9   10

Number of Children

2. How many children chose
   breakfast?

   _____ children

3. Which meal was chosen by
   the most children?

   _____

PROBLEM SOLVING  REAL WORLD

4. What if 10 children chose breakfast?
   How many children could choose lunch or dinner?

   _____ children

Lesson Check

Use the bar graph to answer the question.

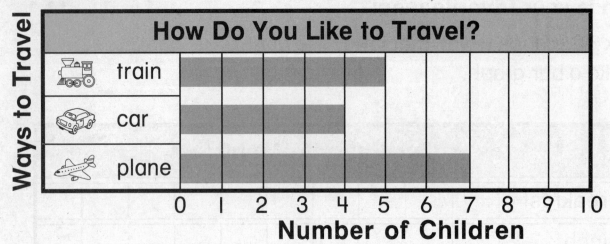

1. How many more children chose ✈ than 🚂?

2   5   7   12

  ○   ○   ○   ○

Spiral Review

2. Which completes the related facts? (Lesson 5.2)

$$7 + 8 = 15 \qquad\qquad 15 - 8 = 7$$
$$8 + 7 = 15$$

  ○ $8 - 7 = 1$ ○ $15 - 7 = 8$

  ○ $7 + 7 = 14$ ○ $8 + 8 = 16$

3. What is the sum? (Lesson 8.7)

$$\begin{array}{r} 43 \\ + 21 \\ \hline \end{array}$$

21	24	62	64
○	○	○	○

Name _____

Read Tally Charts

Complete the tally chart.

Our Favorite Vegetable		Total
beans	IIII	
corn	IIII III	
carrots	IIII	

Use the tally chart to answer each question.

1. How many children chose ? _____ children

....

2. How many children chose ? _____ children

....

3. How many more children chose than ? _____ more children

....

4. Which vegetable did the most children choose? Circle.

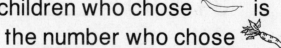

PROBLEM SOLVING REAL WORLD

Complete each sentence about the tally chart.
Write **greater than**, **less than**, or **equal to**.

5. The number of children who chose  is
_____ the number who chose _____ .

6. The number of children who chose _____ is
_____ the number who chose _____ .

Lesson Check

Use the tally chart to answer each question.

Our Favorite Pet	
🐕 dog	☐☐☐☐☐ ‖
🐈 cat	☐☐☐☐☐
🐟 fish	‖

1. How many children chose ?

   ○ 4
   ○ 5
   ○ 7
   ○ 8

2. How many more children chose than ?

   4 more   5 more   6 more   7 more
     ○   ○   ○   ○

Spiral Review

3. There are 8 apples.
   6 apples are red.
   The rest are green.
   How many apples are green? (Lesson 2.4)

   7   5   2   0
   ○   ○   ○   ○

4. What is the sum? (Lesson 8.5)

   $$34 + 40 = \underline{\qquad}$$

   64   70   73   74
   ○   ○   ○   ○

Make Tally Charts

Which color do most children like best? Ask 10 friends. Make 1 tally mark for each child's answer.

Favorite Color		Total
red		
blue		

1. How many children chose red?

   _____ children

2. How many children chose blue?

   _____ children

3. Circle the color that was chosen by fewer children.

   red   blue

PROBLEM SOLVING

Jason asked 10 friends to choose their favorite game. He will ask 10 more children.

Our Favorite Game	
tag	I
kickball	⊦⊦⊦⊦ II
hopscotch	II

4. Predict. Which game will children most likely choose?

   _____

5. Predict. Which game will children least likely choose?

   _____

Lesson Check

1. Which insect did the most children choose?

Our Favorite Insect		Total
ladybug	III	3
bee	I	1
butterfly	~~IIII~~ II	7

○   ○   ○   ○

....

Spiral Review

2. Which number is greater than 54? **(Lesson 7.1)**

45   50   54   57

○   ○   ○   ○

....

3. Which shows the same number? **(Lesson 6.3)**

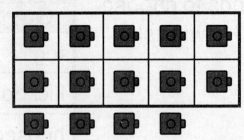

4   10   14   41

○   ○   ○   ○

Problem Solving • Represent Data

Bella made a tally chart to show the favorite sport of 10 friends.

Our Favorite Sport	
Soccer	IIII I
Basketball	III
Baseball	I

Use the tally chart to make a bar graph.

Our Favorite Sport								
Soccer								
Basketball								
Baseball								

Kinds of Sports

0 1 2 3 4 5 6 7 8
Number of Children

Use the graph to solve.

1. How many friends chose soccer?

____ friends

2. How many friends chose soccer or basketball?

____ friends

Lesson Check

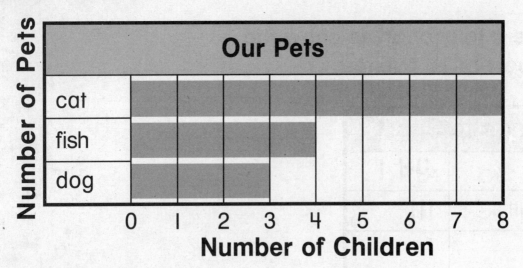

Our Pets

Number of Pets (vertical axis): cat, fish, dog

Number of Children (horizontal axis): 0 1 2 3 4 5 6 7 8

1. Use the graph. How many more children have fish than a dog?

1 more	3 more	4 more	5 more
○	○	○	○

Spiral Review

2. Which ribbon is the shortest? (Lesson 9.1)

○ ▭
○ ▭
○ ▭
○ ▭

3. Look at the hour hand. What is the time? (Lesson 9.6)

6 o'clock	5:00	4:00	3 o'clock
○	○	○	○

Name _____

Chapter 10 Extra Practice

Lessons 10.1 – 10.2 (pp. 413–420) ·

Use the picture graph to answer the question.

What We Read Last Night								
picture book	�	☻	☻	☻	☻			
chapter book	☻	☻	☻	☻				
comic book	☻	☻	☻	☻	☻	☻	☻	

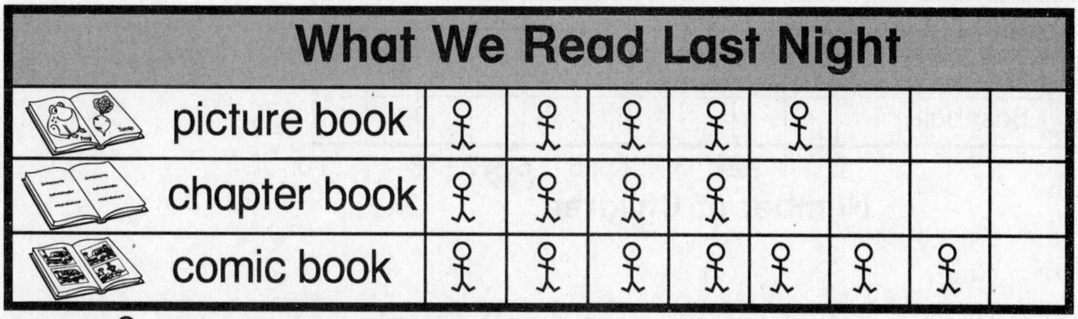

Each  ☻ stands for 1 child.

1. How many children read last night? _____ children

_____

2. Which book did the most children read? Circle.

Lesson 10.3 (pp. 421–424) ·

Use the bar graph to answer the question.

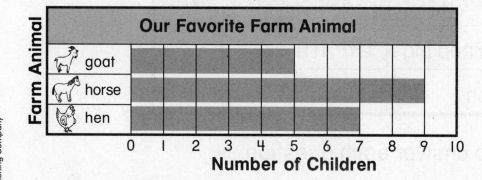

1. How many children chose 🐔?

   _____ children

2. How many more children chose 🐴 than 🐴?

   _____ more children

Lesson 10.4 (pp. 425–427) ·

Ask 10 friends which toy they like best.

1. Make a bar graph.

Our Favorite Toy											
toy rabbit											
puzzle											
baseball											

Toy

0 1 2 3 4 5 6 7 8 9 10
Number of Children

2. Which toy did the most children choose? Circle.

3. How many children chose ? _____ children

Lessons 10.5 – 10.6 (pp. 429–436) ·

Complete the tally chart.

Our Favorite Class Pet		Total
guinea pig	‖‖‖ ‖‖‖‖	
fish	‖‖‖‖	

Use the tally chart to answer each question.

1. Which pet did more children choose? Circle.

_____

2. How many more children chose

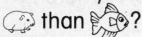 than ? _____ more children

P210 two hundred ten

School-Home
Letter

Dear Family,

My class started Chapter 11 this week. In this chapter, I will learn about three-dimensional shapes. I will learn how to make objects and larger shapes from other shapes.

Love, _____

Vocabulary Builder

flat surface

curved surface

Home Activity

Use a paper towel roll (cylinder), a tennis ball (sphere), a cube-shaped box or building block (cube), and a book (rectangular prism). Build objects using these or other household items of the same shapes. Have children name each shape used in the objects you make.

Literature

Look for these books in a library. Point out shapes and how they can be found in everyday objects.

The Greedy Triangle
Marilyn Burns. Scholastic, 2008.

Captain Invincible and the Space Shapes
Stuart J. Murphy. HarperCollins Publishers, 2001.

Carta
para la casa

Querida familia:

Mi clase comenzó el Capítulo 11 esta semana. En este capítulo, aprenderé sobre las guras tridimensionales. Aprenderé a hacer objetos y guras más grandes tomando como base otras guras.

Con cariño, _____

Vocabulario

superficie plana

superficie curva

Actividad para la casa

Use un rollo de papel (cilindro), una pelota de tenis (esfera), una caja con forma de cubo o un bloque de construcción (cubo) y un libro (prisma rectangular). Construya objetos usando estas u otras cosas con formas similares que encuentre en la casa. Pídale a su hijo que nombre cada figura usada en los objetos que usted haga.

Literatura

Busque estos libros en una biblioteca. Señale las figuras y muestre a su hijo cómo las puede encontrar en los objetos que ve a diario.

The Greedy Triangle
por Marilyn Burns.
Scholastic, 2008.

Captain Invincible and the Space Shapes
por Stuart J. Murphy.
HarperCollins
Publishers, 2001.

Name _____

Three-Dimensional Shapes

Use three-dimensional shapes.
Write the number of flat surfaces for
each shape.

1. A cylinder has ___ flat surfaces.

....

2. A rectangular prism has ___ flat surfaces.

....

3. A cone has ___ flat surface.

....

4. A cube has ___ flat surfaces.

....

PROBLEM SOLVING REAL WORLD

5. Circle the object that matches the clue.
    Mike finds an object that has only a curved surface.

Lesson Check

1. Which shape has both flat and curved surfaces?

 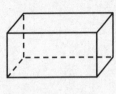

○   ○   ○   ○

2. Which shape has only a curved surface?

○   ○   ○   ○

Spiral Review

3. Count forward. What number is missing? (Lesson 6.1)

109, 110, 111, ___, 113

107   108   112   117

○   ○   ○   ○

4. What is the sum of 2 and 3? (Lesson 1.2)

1   4   5   6

○   ○   ○   ○

Name _____

Combine Three-Dimensional Shapes

Use three-dimensional shapes.

Combine.	**Which new shape can you make? Circle it.**
1.	
2.	
3.	

PROBLEM SOLVING REAL WORLD

4. Circle the shapes you could use to model the bird feeder.

Lesson Check

1. Which shape combines and ?

○   ○   ○   ○

Spiral Review

2. What is the sum? (Lesson 8.2)

$$40 + 20 = \underline{\hspace{1cm}}$$

60   50   42   20
○   ○   ○   ○

3. Emi has 15 crayons. She gives
   some crayons to Jo. Now she
   has 9 crayons. How many crayons
   did Emi give to Jo? (Lesson 5.1)

4   6   7   12
○   ○   ○   ○

Make New Three-Dimensional Shapes

Use three-dimensional shapes.

Build and Repeat.	Combine. Which new shape can you make? Circle it.
1.	
2.	
3.	

PROBLEM SOLVING REAL WORLD

4. Dave builds this shape.
   Then he repeats and combines.
   Draw a shape he can make.

Lesson Check

1. Which new shape can you make?

 Combine and .

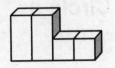

○   ○   ○   ○

···

Spiral Review

2. Which addition fact helps you solve $15 - 6 =$ ____? (Lesson 4.3)

$$6 + 5 = 11$$
○

$$8 + 6 = 14$$
○

$$6 + 7 = 13$$
○

$$6 + 9 = 15$$
○

···

3. Which doubles fact helps you solve $5 + 6 = 11$? (Lesson 3.5)

$$3 + 3 = 6$$
○

$$4 + 4 = 8$$
○

$$5 + 5 = 10$$
○

$$7 + 7 = 14$$
○

Name _____

Problem Solving • Take Apart Three-Dimensional Shapes

Use three-dimensional shapes.
Circle your answer.

....

1. Paco used shapes to build this robot. Circle the shapes he used.

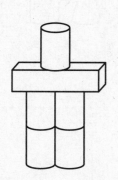   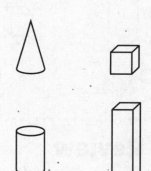

....

2. Eva used shapes to build this wall. Circle the shapes she used.

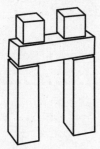   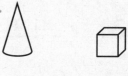 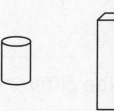

....

PROBLEM SOLVING REAL WORLD

3. Circle the ways that show the same shape.

   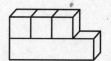

Lesson Check

1. Which shapes are used to make the picture frame?

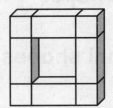

  ◯

  ◯

  ◯

  ◯

....

Spiral Review

2. Which is true? (Lesson 7.3)

$13 > 31$ $13 = 31$ $31 < 13$ $31 > 13$

◯   ◯   ◯   ◯

....

3. What is the difference? (Lesson 8.9)

$$60 - 30 = \underline{\quad\quad}$$

60   30   20   3

◯   ◯   ◯   ◯

Name _____

Two-Dimensional Shapes on Three-Dimensional Shapes

Circle the objects you could trace to draw the shape.

1.

2.

3.

PROBLEM SOLVING · REAL WORLD

4. Look at this shape. Draw the shape you would make if you traced this object.

Lesson Check

1. Which flat surface does a cone have?

○   ○   ○   ○

2. Which flat surfaces could a rectangular prism have?

○   ○   ○   ○

Spiral Review

3. Jade has 8 books. She gives some
   of them to Dana. Now Jade has 6 books.
   How many did she give to Dana? Which
   subtraction sentence answers the problem? (Lesson 4.1)

   $9 - 3 = 6$ $9 - 2 = 7$ $8 - 3 = 5$ $8 - 2 = 6$

   ○   ○   ○   ○

4. What is the sum? (Lesson 1.5)

   $3 + 0 = \underline{\hspace{1cm}}$

   0   1   3   4

   ○   ○   ○   ○

Chapter 11 Extra Practice

Lesson 11.1 (pp. 457–460) . . . .

Use three-dimensional shapes.
Write the number of flat surfaces
for each shape.

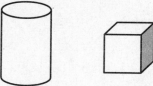

1. A cone has __ flat surfaces.

2. A cube has __ flat surfaces.

3. A cylinder has __ flat surfaces.

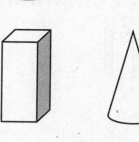

4. A rectangular prism has __ flat surfaces.

Lesson 11.2 (pp. 461–464) . . . .

Use three-dimensional shapes.

Combine.	Which new shape can you make? Circle it.

1.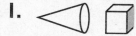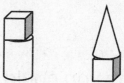

2.

Lesson 11.3 (pp. 465–467)

Use three-dimensional shapes.

Build and Repeat.	**Combine. Which new shape can you make? Circle it.**
1.	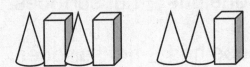
2.	

Lesson 11.5 (pp. 473–476)

Circle the objects you could trace to draw the shape.

1.

2.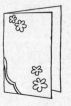

School-Home
Letter

Dear Family,

My class started Chapter 12 this week. In this chapter, I will describe and combine two-dimensional shapes. I will learn about equal shares, halves, and fourths.

Love, _____

Vocabulary Builder

hexagon

trapezoid

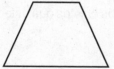

Home Activity

Use a napkin (square), a folded napkin (triangle), and an envelope (rectangle). Combine these items or other household items of the same shapes to make new shapes. Have your child name each shape used in the new shapes you made.

Literature

Look for these books in a library. Point out shapes and how they can be found in everyday objects.

The Greedy Triangle by Marilyn Burns. Scholastic, 2008.

Color Farm by Lois Ehlert. HarperCollins, 1990.

Capítulo 12

Carta para la casa

Querida familia:

Mi clase comenzó el Capítulo 12 esta semana. En este capítulo, aprenderé sobre figuras bidimensionales. Aprenderé cómo hacer figuras más grandes que otras.

Con cariño, _____

Vocabulario

hexágono

trapecio

Actividad para la casa

Use una servilleta (cuadrado), una servilleta doblada (triángulo) y un sobre (rectángulo). Construya objetos usando estos u otros elementos de la casa con las mismas formas. Pídales a los niños que nombren cada figura usada en los objetos que usted hace.

Literatura

Busque estos libros en una biblioteca. Señale las figuras y muestre cómo se pueden encontrar en los objetos de la vida diaria.

The Greedy Triangle
por Marilyn Burns. Scholastic, 2008.

Color Farm
by Lois Ehlert. HarperCollins, 1990.

Name _____

Sort Two-Dimensional Shapes

Read the sorting rule. Circle the shapes that follow the rule.

1. not curved

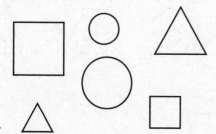

2. 4 vertices

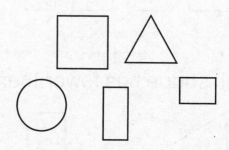

3. more than 3 sides

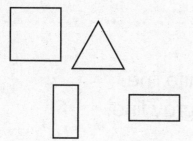

4. curved

PROBLEM SOLVING REAL WORLD

5. Katie sorted these shapes. Write a sorting rule to tell how Katie sorted.

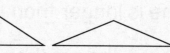

_____

- - - - - - - - - - - - - - - -

_____

Lesson Check

1. Which shape would **not** be sorted into this group?

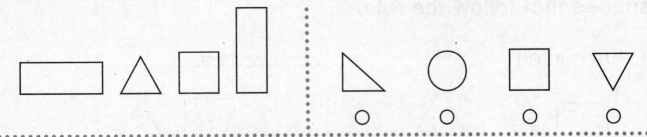

2. Which shape has fewer than 4 sides?

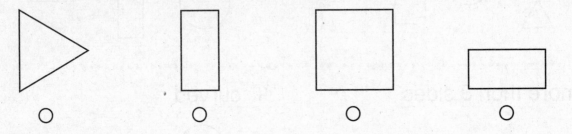

Spiral Review

3. Clue 1: A black line is shorter than a white line.
Clue 2: The white line is shorter than a gray line.
Use the clues. Which is true? (Lesson 9.2)

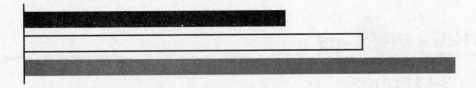

○ The black line is longer than the gray line.

○ The black line is shorter than the gray line.

○ The white line is shorter than the black line.

○ The white line is longer than the gray line.

Describe Two-Dimensional Shapes

Use 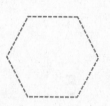 to trace each straight side. Use to circle each vertex. Write the number of sides and vertices.

1.

_____ sides

_____ vertices

2.

_____ sides

_____ vertices

3.

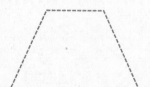

_____ sides

_____ vertices

4.

_____ sides

_____ vertices

5.

_____ sides

_____ vertices

6.

_____ sides

_____ vertices

PROBLEM SOLVING REAL WORLD

Draw a shape to match the clues.

7. Ying draws a shape with 4 sides. She labels it as a rectangle.

Lesson Check

1. How many vertices does a triangle have?

3	4	5	6
○	○	○	○

2. How many vertices does a ▢ have?

2	3	4	5
○	○	○	○

Spiral Review

3. Count on to find 2 + 9. (Lesson 3.2)

15	13	11	5
○	○	○	○

4. Corey measures a crayon box with his paper clip ruler. About how long is the box? (Lesson 9.5)

   ○ about 2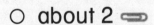

   ○ about 3 ⬭

   ○ about 4 ⬭

   ○ about 5 ⬭

Combine Two-Dimensional Shapes

Use pattern blocks. Draw to show the blocks. Write how many blocks you used.

1. How many △ make a ⬠?

2. How many △ make a ◇?

_____ △ make a ⬠.

_____ △ make a ◇.

PROBLEM SOLVING REAL WORLD

Use pattern blocks. Draw to show your answer.

3. 2 ⬠ make a ⬡.

How many ⬠ make 4 ⬡?

_____ ⬠ make 4 ⬡.

Lesson Check

1. How many △ do you use to make a ⬡?

6	5	4	3
○	○	○	○

2. How many ◇ do you use to make a ⬡?

6	5	3	2
○	○	○	○

Spiral Review

3. Use 🖇. Which string is about 5 🖇 long? (Lesson 9.4)

○

○

○

○

4. Look at the hour hand. What is the time? (Lesson 9.7)

○ half past 4:00

○ half past 5:00

○ 5:00

○ 4:00

Combine More Shapes

**Circle two shapes that can combine
to make the shape on the left.**

1.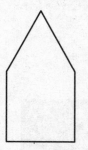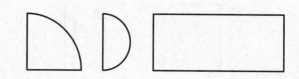

2.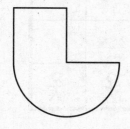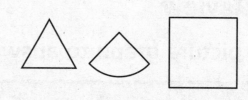

3.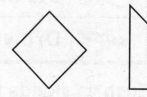

PROBLEM SOLVING REAL WORLD

4. Draw lines to show how the shapes
   on the left combine to make the new shape.

   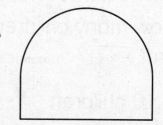

Lesson Check

1. Which shapes can combine to make this new shape?

 ○    ○    ○    ○

Spiral Review

Use the picture graph to answer each question.

Our Favorite Activity						
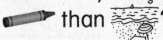 Swimming	웃	웃	웃			
Dancing	웃	웃	웃	웃		
Drawing	웃	웃	웃	웃	웃	웃

Each 웃 stands for 1 child.

2. How many more children chose than ? (Lesson 10.1)

6 children ○   4 children ○   3 children ○   2 children ○

3. How many children chose and ? (Lesson 10.2)

10 children ○   9 children ○   7 children ○   4 children ○

Problem Solving • Make New Two-Dimensional Shapes

Use shapes to solve.
Draw to show your work.

1. Use ⬜ to make a ▭.
   Step 1. Combine shapes to make a new shape.

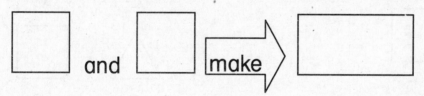

Step 2. Then use the new shape.

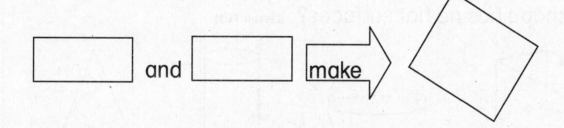

....

2. Use ◿ to make a ◯.
   Step 1. Combine shapes to make a new shape.

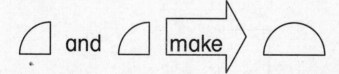

Step 2. Then use the new shape.

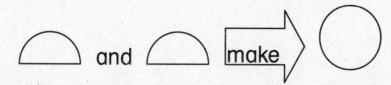

Lesson Check

1. Which new shape could you make?

Step 1.
Combine 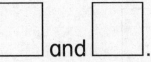 and to make .

Step 2.
Then use 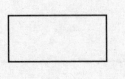 and .

Spiral Review

2. Which shape has no flat surfaces? **(Lesson 11.1)**

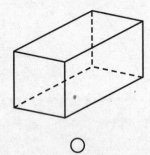 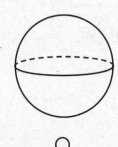 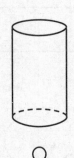 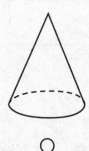

○   ○   ○   ○

3. Which flat surface does a cylinder have? **(Lesson 11.5)**

○   ○   ○   ○

Find Shapes in Shapes

Use two pattern blocks to make the shape.
Draw a line to show your model. Circle the blocks you use.

1.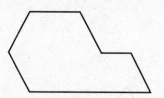

2.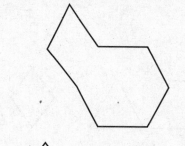

3.

4.

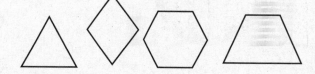

PROBLEM SOLVING REAL WORLD

Make the shape to the right. Use the number
of pattern blocks listed in the exercise.
Write how many of each block you use.

5. Use 3 blocks.

Lesson Check

1. Which two pattern blocks can make this shape?

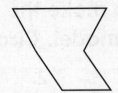

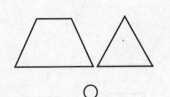 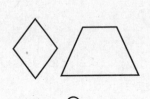 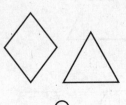 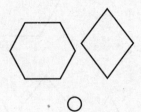

○   ○   ○   ○

Spiral Review

2. What time is it? **(Lesson 9.8)**

○ 2:00

○ 2:30

○ 3:00

○ 3:30

3. Which tally marks show the number 8? **(Lesson 10.5)**

ℍℍ ℍℍ ||| ℍℍ ℍℍ   ℍℍ ||| |||

○   ○   ○   ○

4. How many vertices does a ☐ have? **(Lesson 12.2)**

4   3   2   1

○   ○   ○   ○

Take Apart Two-Dimensional Shapes

Draw a line to show the parts.

1. Show 2 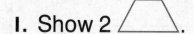.

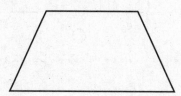

2. Show 2 △.

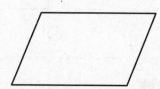

3. Show 1 ☐ and 1 ▭.

4. Show 1 ⬡ and 1 △.

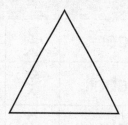

PROBLEM SOLVING REAL WORLD

5. How many triangles are there?

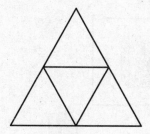

_____ triangles

Lesson Check

1. Look at the picture.
What are the parts?

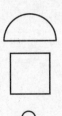

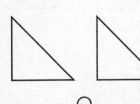 ○

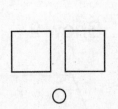 ○

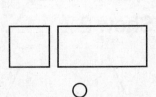 ○

○

Spiral Review

2. Use the graph.
How many children chose ? (Lesson 10.1)

Our Favorite Sport							
🔵 soccer	⚲	⚲					
⚾ baseball	⚲	⚲	⚲	⚲	⚲	⚲	
🎾 tennis	⚲	⚲	⚲	⚲	⚲		

Each ⚲ stands for 1 child.

7 children ○   6 children ○   5 children ○   2 children ○

3. Which new shape
can you make? (Lesson 11.3)

Combine 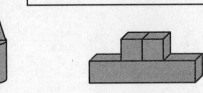 and .

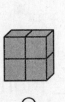 ○

 ○

○

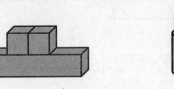 ○

Name _____

Equal or Unequal Parts

Color the shapes that show unequal shares.

1.

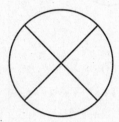 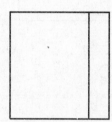 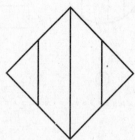 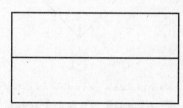

Color the shapes that show equal shares.

2.

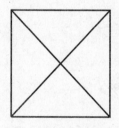 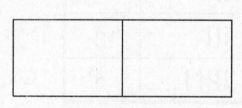 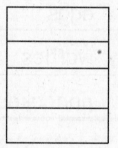 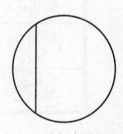

PROBLEM SOLVING REAL WORLD

Draw lines to show the parts.

3. 4 equal shares

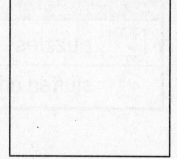

Lesson Check

1. Which shows unequal shares?

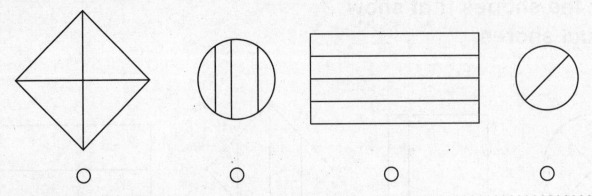

○　　　　　○　　　　　○　　　　　○

Spiral Review

2. Which food did the most children choose? (Lesson 10.6)

Our Favorite Breakfast		Total
eggs	||||	4
waffles	|||	3
pancakes	⊮|	6

○

○

○

○

3. Use the graph. How many children chose 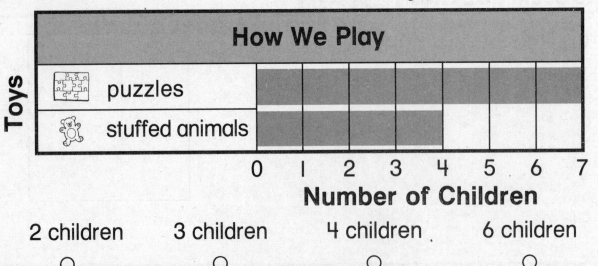 ? (Lesson 10.3)

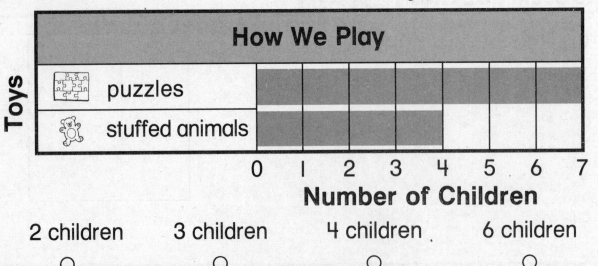

2 children　　3 children　　4 children　　6 children

○　　　　　○　　　　　○　　　　　○

Halves

Circle the shapes that show halves.

1.

2.

3.

4.

5.

6.

7.

8.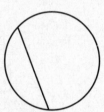

9.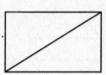

PROBLEM SOLVING REAL WORLD

Draw or write to solve.

10. Kate cut a square into equal shares. She traced one of the parts. Write **half of** or **halves** to name the part.

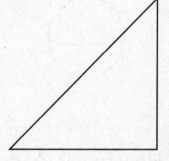

_____

_ _ _ _ _ _ _ _ _ _ _

_____ a square

Lesson Check

1. Which shows halves?

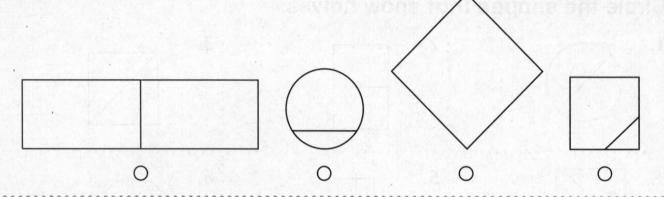

$\bigcirc$ $\bigcirc$ $\bigcirc$ $\bigcirc$

Spiral Review

2. Which new shape can you make? (Lesson 11.2)

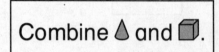

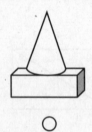      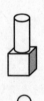

$\bigcirc$ $\bigcirc$ $\bigcirc$ $\bigcirc$

3. Which shape has both flat and curved surfaces? (Lesson 11.1)

$\bigcirc$ $\bigcirc$ $\bigcirc$ $\bigcirc$

4. How many $\triangle$ do you use to make a $\triangle$? (Lesson 12.3)

8   6   4   3

$\bigcirc$ $\bigcirc$ $\bigcirc$ $\bigcirc$

Fourths

Circle the shapes that show fourths.

1.

2.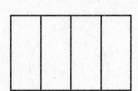

3.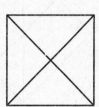

4.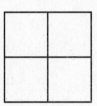

5.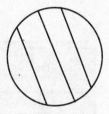

6.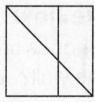

7.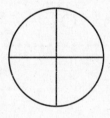

8.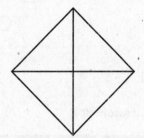

9.

PROBLEM SOLVING REAL WORLD

Solve.

10. Chad drew a picture to show a quarter of a circle. Which shape did Chad draw? Circle it.

Lesson Check

1. Which shows fourths?

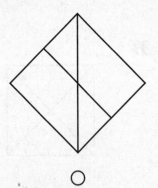 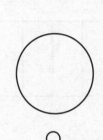 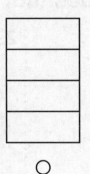 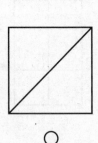

○　　　　　　　○　　　　　　　○　　　　　　　○

Spiral Review

2. What shapes are used to build the wall? (Lesson 11.4)

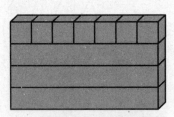

  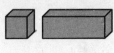

○　　　　　　　○　　　　　　　○　　　　　　　○

3. How many fewer children answered **yes** than **no**? (Lesson 10.4)

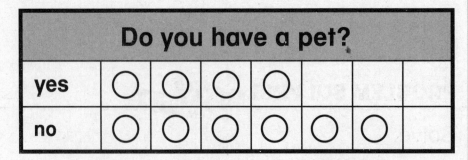

2 fewer children ○

3 fewer children ○

4 fewer children ○

6 fewer children ○

Chapter 12 Extra Practice

Lessons 12.1 – 12.2 . . . .

Use  to trace each straight side.

Use to circle each vertex.

Write the number of sides and vertices.

1. _____ sides

_____ vertices

2. _____ sides

_____ vertices

Lessons 12.3 – 12.4 . . . .

Circle the two shapes that can combine
to make the shape on the left.

1.

Lesson 12.6 . . . .

Use two pattern blocks to make the shape.

Draw a line to show your model.

Circle the blocks you use.

1.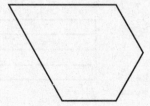

 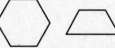

2.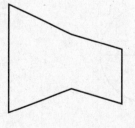

  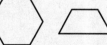

Lesson 12.7
Draw a line to show the parts.

1. Show 2 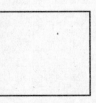.

2. Show 2 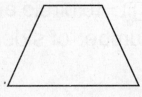.

3. Show 2 .

4. Show 3 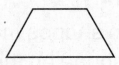.

Lesson 12.8
Color the shapes that show unequal shares.

1.

Lessons 12.9 – 12.10
Circle the shapes that show fourths.

1.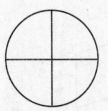

2.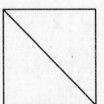

3.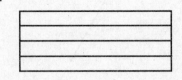

Algebra • Ways to Expand Numbers

Essential Question How can you write a two-digit number in different ways?

Model and Draw

There are different ways to think about a number.

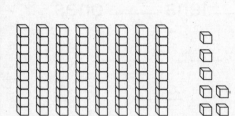

___ tens _7_ ones

80 + 7
___   ___

87
___

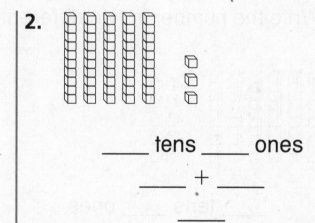
8 tens and 7 ones is the same as 80 plus 7.

Share and Show

Write how many tens and ones.
Write the number in two different ways.

1.

___ tens ___ ones

___ + ___

___

2.

___ tens ___ ones

___ + ___

___

Math Talk   Does the 7 in this number show 7 or 70? Explain.

72

On Your Own

Write how many tens and ones.
Write the number in two different ways.

3.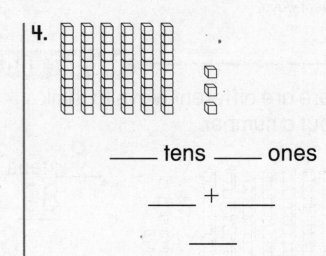

_____ tens _____ ones

_____ + _____

_____

4.

_____ tens _____ ones

_____ + _____

_____

PROBLEM SOLVING

5. Draw the same number using only tens.
Write how many tens and ones.
Write the number in two different ways.

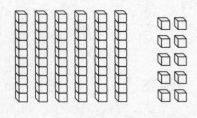

_____ tens _____ ones

_____ + _____

_____

_____ tens _____ ones

_____ + _____

_____

TAKE HOME ACTIVITY • Write a two-digit number to 99.
Ask your child to write how many tens and ones and then write the
number a different way.

Name _____

Identify Place Value

Essential Question How can you use place value to understand the value of a number?

Model and Draw

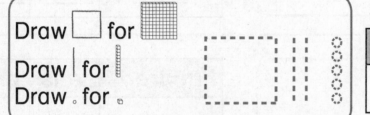

The **1** in **1**25 means 1 hundred.
The **2** in 1**2**5 means 2 tens.
The **5** in 12**5** means 5 ones.

125

Draw ☐ for ▦
Draw | for |
Draw ° for ₀

hundreds	tens	ones
1	2	5

Share and Show

Math Board

Use your MathBoard and ▦ ▭ ▯
to show the number.
Draw to complete the quick picture. Write how many hundreds, tens, and ones.

THINK
106 has no tens.

1.

106

hundreds	tens	ones
___	___	___

Math Talk How is the 1 in 187 different from the 1 in 781?

Getting Ready for Grade 2

two hundred fifty-one **P251**

On Your Own

Use your MathBoard and ▦ ▭ ▫.
Draw to complete the quick picture.
Write how many hundreds, tens, and ones.

2.

170

hundreds	tens	ones
___	___	___

3.

143

hundreds	tens	ones
___	___	___

4.

121

hundreds	tens	ones
___	___	___

PROBLEM SOLVING

Circle your answer.

5. I have 1 hundred, 9 tens, and 9 ones. What number am I?

99   100   199

6. I have 3 ones, 0 tens, and 1 hundred. What number am I?

107   170   103

TAKE HOME ACTIVITY • Write some numbers from 100 to 199. Have your child tell how many hundreds, tens, and ones are in the number.

Name _____

Use Place Value to Compare Numbers

Essential Question How can you use place value to compare two numbers?

Model and Draw

I want to eat the greater number.

Use these symbols to compare numbers.

> is greater than
< is less than
= is equal to

 45 46

45 < 46
45 is less than 46.

Compare 134 and 125.

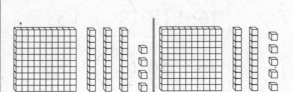

First compare hundreds.
One hundred is equal to one hundred.
100 = 100
If the hundreds are equal, compare the tens. 30 is greater than 20.
134 > 125

Share and Show

Math Board

Write the numbers and compare. Write >, <, or =.

1.  159 ⟩ 155

2. ____ ◯ ____

Compare the numbers using >, <, or =.

3. 187 ◯ 168   4. 165 ◯ 159   5. 127 ◯ 141

 **Math Talk**   Compare 173 and 177. Did you have to compare all the digits? Why or why not?

On Your Own

Write the numbers. Compare. Write >, <, or =.

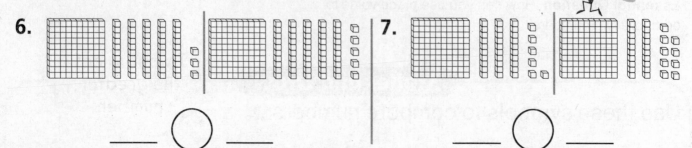

6. ___ ⃝ ___

7. ___ ⃝ ___

Compare the numbers using >, <, or =.

8. 143 ⃝ 143   9. 162 ⃝ 157   10. 185 ⃝ 188

11. 124 ⃝ 129   12. 189 ⃝ 195   13. 135 ⃝ 135

14. 173 ⃝ 164   15. 123 ⃝ 117   16. 118 ⃝ 131

17. 155 ⃝ 145   18. 181 ⃝ 181   19. 192 ⃝ 179

20. 122 ⃝ 129   21. 166 ⃝ 177   22. 154 ⃝ 154

PROBLEM SOLVING   REAL WORLD

23. Antonio is thinking of a number between 100 and 199. It has 1 hundred, 3 tens, and 6 ones. Kim is thinking of a number between 100 and 199. It has 1 hundred, 6 tens, and 3 ones. Who is thinking of a greater number?

Draw or write to explain.

_____ is thinking of a greater number.

TAKE HOME ACTIVITY • Choose two numbers between 100 and 199 and have your child explain which number is greater.

Name _____

Concepts and Skills

Write how many tens and ones.
Write the number in two ways.

1.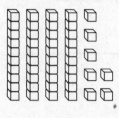

   _____ tens and _____ ones

   _____ + _____

   _____

2.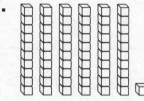

   _____ tens and _____ one

   _____ + _____

   _____

Use your MathBoard and ▦ ◻.
Draw to complete the quick picture.
Write how many hundreds, tens, and ones.

3. 154

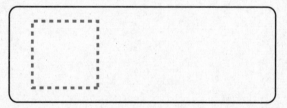

hundreds	tens	ones
_____	_____	_____

4. 128

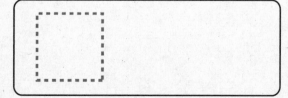

hundreds	tens	ones
_____	_____	_____

Write the numbers and compare. Write >, <, or =.

5.

___ ◯ ___

6.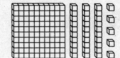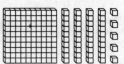

___ ◯ ___

Compare the numbers using >, <, or =.

7. 175 ◯ 175

9. 189 ◯ 188

11. 157 ◯ 157

8. 163 ◯ 173

10. 142 ◯ 158

12. 185 ◯ 180

13. Which comparison is correct?

○ 132 > 131

○ 131 = 132

○ 131 > 132

Algebra • Addition Function Tables

Essential Question How can you follow a rule to complete an addition function table?

Model and Draw

The rule is Add 9. Add 9 to each number.

Add 9	
7	16
8	17
9	18

Share and Show

Follow a rule to complete the table.

1.
Add 3	
7	
8	
9	

2.
Add 4	
6	
7	
8	

3.
Add 5	
5	
7	
9	

4.
Add 8	
5	
7	
9	

5.
Add 7	
6	
8	
9	

6.
Add 6	
6	
8	
9	

Math Talk Look at Exercise 4. How does the rule help you see a pattern?

On Your Own

Follow a rule to complete the table.

7.

Add 7	
7	
8	
9	

8.

Add 4	
7	
8	
9	

9.

Add 5	
7	
8	
9	

10.

Add 8	
4	
6	
8	
9	

11.

Add 3	
3	
5	
7	
9	

12.

Add 6	
6	
7	
8	
9	

PROBLEM SOLVING REAL WORLD

13. Solve. Complete the table.

Tom is 8 years old.
Julie is 7 years old.
Carla is 4 years old.

How old will each child
be in 4 years?

Tom	8	
Julie	7	
Carla	4	

TAKE HOME ACTIVITY • Copy Exercise 12 and change the numbers in the
left column to 9, 7, 5, and 3. Have your child complete the table and
explain how he or she used a rule to solve the problem.

Algebra • Subtraction Function Tables

Essential Question How can you follow a rule to complete a subtraction function table?

Model and Draw

The rule is Subtract 7. Subtract 7 from each number.

Subtract 7	
14	7
15	8
16	9

Share and Show

Follow a rule to complete the table.

1.

Subtract 3	
9	
10	
11	

2.

Subtract 4	
6	
8	
10	

3.

Subtract 5	
6	
8	
10	

4.

Subtract 8	
9	
11	
13	

5.

Subtract 7	
12	
13	
14	

6.

Subtract 6	
6	
8	
9	

Math Talk How can Exercise 2 help you solve Exercise 3?

On Your Own

Follow a rule to complete the table.

7.

Subtract 4	
11	
12	
13	

8.

Subtract 6	
7	
8	
9	

9.

Subtract 5	
7	
8	
9	

10.

Subtract 7	
13	
14	
15	
16	

11.

Subtract 8	
12	
14	
16	
17	

12.

Subtract 9	
12	
14	
16	
17	

PROBLEM SOLVING REAL WORLD

13. Solve. Complete the table.

Jane has 4 cookies.
Lucy has 3 cookies.
Seamus has 2 cookies.

How many cookies will each child have if they each eat 2 cookies?

Jane	4	
Lucy	3	
Seamus	2	

TAKE HOME ACTIVITY • Copy Exercise 12 and change the numbers in the left column to 10, 11, 12, and 13. Have your child complete the table and explain how he or she used a rule to solve the problem.

Algebra • Follow the Rule

Essential Question How can you follow a rule to complete an addition or subtraction function table?

Model and Draw

The rule for some tables is to add. For other tables the rule is to subtract.

Add 1	
2	3
4	
6	
8	

Subtract 1	
2	1
4	
6	
8	

Share and Show

Follow a rule to complete the table.

1.

Add 2	
10	
9	
8	
7	

2.

Subtract 2	
10	
9	
8	
7	

3.

Subtract 1	
3	
4	
7	
9	

Math Talk What is the rule for the pattern in
Exercise 1?

On Your Own

Follow a rule to complete the table.

4.

Add 5	
7	
8	
9	
10	

5.

Subtract 5	
7	
8	
9	
10	

6.

Subtract 1	
8	
9	
11	
13	

7.

Subtract 3	
5	
7	
9	
11	

8.

Add 4	
6	
7	
8	
9	

9.

Add 6	
9	
8	
7	
6	

PROBLEM SOLVING

10. Find the rule. Complete the table.

3	
	8
7	10
	12

TAKE HOME ACTIVITY • Copy the table for Exercise 9.
Change the rule to Subtract 3. Have your child complete the table.

P262 two hundred sixty-two

Name _____

Add 3 Numbers

Essential Question How can you choose a strategy to help add 3 numbers?

Model and Draw

When you add 3 numbers, you can add in any order.
Using a strategy can help.

Make a 10.	**Use doubles.**	**Use count on.**
$\begin{array}{r} 2 \\ 6 \\ +\,8 \\ \hline 16 \end{array}$ 〉10 +6	$\begin{array}{r} 8 \\ 8 \\ +\,4 \\ \hline 20 \end{array}$ 〉16 +4	$\begin{array}{r} 6 \\ 8 \\ +\,3 \\ \hline 17 \end{array}$ 〉9 +8

Share and Show

Use strategies to find the sums. Circle any strategy you use.

1. $\begin{array}{r} 4 \\ 7 \\ +\,7 \\ \hline \end{array}$ make a 10
    doubles
    count on

2. $\begin{array}{r} 9 \\ 8 \\ +\,1 \\ \hline \end{array}$ make a 10
    doubles
    count on

3. $\begin{array}{r} 4 \\ 6 \\ +\,2 \\ \hline \end{array}$ make a 10
    doubles
    count on

4. $\begin{array}{r} 8 \\ 4 \\ +\,2 \\ \hline \end{array}$ make a 10
    doubles
    count on

5. $\begin{array}{r} 6 \\ 3 \\ +\,6 \\ \hline \end{array}$ make a 10
    doubles
    count on

6. $\begin{array}{r} 6 \\ 7 \\ +\,4 \\ \hline \end{array}$ make a 10
    doubles
    count on

Math Talk Explain why you used the make a 10 strategy to solve Exercise 2.

On Your Own

Use a strategy to find the sum. Circle the strategy you choose.

7.   5 make a 10
     5 doubles
  + 5 count on
	8. 7 make a 10
	     3 doubles
	  + 5 count on
		9. 3 make a 10
		     8 doubles
		  + 8 count on

10.   4 make a 10
      2 doubles
   + 7 count on
	11. 2 make a 10
	     9 doubles
	  + 2 count on
		12. 9 make a 10
		      9 doubles
		   + 1 count on

13.   9 make a 10
      2 doubles
   + 8 count on
	14. 6 make a 10
	     3 doubles
	  + 7 count on
		15. 8 make a 10
		     4 doubles
		  + 1 count on

PROBLEM SOLVING REAL WORLD

16. Christine has 7 red buttons, 3 blue buttons, and 4 yellow buttons. How many buttons does she have?

_____ buttons

TAKE HOME ACTIVITY • Ask your child to choose 3 numbers from 1 to 9. Have your child add to find the sum.

Write the sum.

	14.	15.	16.
22 + 7	53 + 3	46 + 2	71 + 8
84 + 5	18. 93 + 4	19. 16 + 3	20. 37 + 1
62 + 2	22. 23 + 5	23. 82 + 2	24. 44 + 4

ROBLEM SOLVING REAL WORLD

There are 23 children in the first grade class. Then 3 more children join the class. How many children are there now?

_____ children

© Houghton Mifflin Harcourt Publishing Company

TAKE HOME ACTIVITY • Tell your child you had 12 pennies and then you got 5 more. Have your child add to find how many pennies in all.

two hundred sixty-six

L

Add a One-Digit Number to a Two-Digit Number

Add.

Essential Question How can you find the sum of a 1-digit number and a 2-digi

Model and Draw

What is 54 + 2?

To find the sum, find how many **tens** and

	5 tens	4 ones		5 4
+		2 ones		+ 2
	5 tens	6 ones		5 6

13.

17.

21.

Share and Show

Add. Write the sum.

1.	72	2.	24	3.	41
	+ 3		+ 1		+ 4

5.	14	6.	33	7.	61
	+ 4		+ 6		+ 8

9.	31	10.	11	11.	40
	+ 6		+ 7		+ 4

P

25.

Math Talk How did you find the total number of one
Exercise 1?

© Houghton Mifflin Harcourt Publishing Company

Name _____

Add Two-Digit Numbers

Essential Question How can you find the sum of two 2-digit numbers?

Model and Draw

What is 23 + 14?

You can find how many **tens** and **ones** in all.

2	tens	**3** ones		**2 3**
+ **1**	ten	**4** ones		+ **1 4**
__3__	tens	__7__ ones		37

Share and Show

Add. Write the sum.

1.	82 + 12	2.	25 + 43	3.	15 + 14	4.	71 + 12
5.	36 + 21	6.	43 + 41	7.	57 + 32	8.	21 + 12
9.	12 + 12	10.	41 + 21	11.	32 + 41	12.	51 + 14

Math Talk   How many tens are in 26 + 11?
How do you know?

Getting Ready for Grade 2

On Your Own

Add. Write the sum.

13. 83
   + 12

14. 73
   + 21

15. 16
   + 51

16. 23
   + 43

17. 24
   + 55

18. 67
   + 21

19. 64
   + 23

20. 51
   + 24

21. 26
   + 32

22. 51
   + 25

23. 46
   + 22

24. 34
   + 45

PROBLEM SOLVING REAL WORLD

25. Emma has 21 hair clips. Her sister has 11 hair clips. How many hair clips do the girls have together?

_____ hair clips

TAKE HOME ACTIVITY • Tell your child you drove 21 miles and then you drove 16 more. Have your child add to find how many miles in all.

Repeated Addition

Essential Question How can you find how many items there are in equal groups without counting one at a time?

Model and Draw

When all groups have the same number they are equal groups.

Ayita is putting 2 plants on each step up to her porch. She has 4 steps. How many plants does she need?

There are 4 equal groups. There are 2 in each group. Add to find how many in all.

$$\underline{2} + \underline{2} + \underline{2} + \underline{2} = \underline{8}$$

Ayita needs __8__ plants.

Share and Show

Use your MathBoard and ⬤. Make equal groups. Complete the addition sentence.

	Number of Equal Groups	Number in Each Group	How many in all?
1.	4	3	___ + ___ + ___ + ___ = ___
2.	2	5	___ + ___ = ___
3.	3	4	___ + ___ + ___ = ___

Math Talk How can you use addition to find 5 groups of 4?

On Your Own

Use your MathBoard and ⬤. Make equal groups. Complete the addition sentence.

	Number of Equal Groups	Number in Each Group	How many in all?
4.	2	3	___ + ___ = ___
5.	3	5	___ + ___ + ___ = ___
6.	4	4	___ + ___ + ___ + ___ = ___
7.	4	5	___ + ___ + ___ + ___ = ___
8.	5	7	___ + ___ + ___ + ___ + ___ = ___

PROBLEM SOLVING REAL WORLD

Solve.

9. There are 3 flower pots. There are 2 flowers in each flower pot. How many flowers are there?

    ____ flowers

10. There are 2 plants. There are 4 leaves on each plant. How many leaves are there?

    ____ leaves

TAKE HOME ACTIVITY • Use dry cereal or pasta to make 3 equal groups of 5. Ask your child to find the total number of items.

Name _____

Use Repeated Addition to Solve Problems

Essential Question How can you use repeated addition to solve problems?

Model and Draw

Dyanna will have 3 friends at her party. She wants to give each friend 4 balloons. How many balloons does Dyanna need?

THINK 4 + 4 + 4 = 12

__12__ balloons

Share and Show

Draw pictures to show the story.
Write the addition sentence to solve.

1. Ted plays with 2 friends. He wants to give each friend 5 cards. How many cards does Ted need?

_____ cards

2. Aisha shops with 4 friends. She wants to buy each friend 2 roses. How many roses does Aisha need?

_____ roses

Math Talk What pattern can you use to find the answer to Exercise 2?

On Your Own

Draw pictures to show the story.
Write the addition sentence to solve.

3. Lea plays with 3 friends. She wants
   to give each friend 5 ribbons. How
   many ribbons does Lea need?

   _____ ribbons

4. Harry shops with 5 friends. He wants
   to buy each friend 2 pens. How
   many pens does Harry need?

   _____ pens

5. Cam plays with 4 friends. She wants
   to give each friend 4 stickers. How
   many stickers does Cam need?

   _____ stickers

PROBLEM SOLVING  REAL WORLD

Circle the way you can model the problem.
Then solve.

6. There are 4 friends. Each
   friend has 3 apples. How
   many apples are there?

   4 groups of 4 apples
   4 groups of 3 apples
   3 groups of 4 apples

   There are _____ apples.

TAKE HOME ACTIVITY • Use small items such as cereal pieces to act out each problem. Have your child check the answers on this page.

Checkpoint

Concepts and Skills

Follow the rule to complete each table.

1.

Add 3	
2	
4	
6	
8	

2.

Subtract 7	
10	
12	
13	
14	

3.

Add 6	
10	
9	
8	
7	

4.

Subtract 6	
15	
14	
13	
12	

Use strategies to find the sums. Circle any strategy you use.

5. $\begin{array}{r} 4 \\ 3 \\ +4 \\ \hline \end{array}$ make a 10
    doubles
    count on

6. $\begin{array}{r} 3 \\ 7 \\ +5 \\ \hline \end{array}$ make a 10
    doubles
    count on

Add. Write the sum.

7. $\begin{array}{r} 32 \\ +14 \\ \hline \end{array}$

8. $\begin{array}{r} 52 \\ +46 \\ \hline \end{array}$

9. $\begin{array}{r} 18 \\ +21 \\ \hline \end{array}$

10. $\begin{array}{r} 43 \\ +35 \\ \hline \end{array}$

Use your MathBoard and ⬤. Make equal groups.
Complete the addition sentence.

	Number of Equal Groups	Number in Each Group	How many in all?
11.	3	2	___ + ___ + ___ = ___
12.	2	4	___ + ___ = ___

13. Choose the way to model the problem.
    James has 4 letters. He puts 2 stamps on each letter.
    How many stamps does he use in all?

    ○ 2 groups of 4 stamps   ○ 4 groups of 4 stamps

    ○ 2 groups of 2 stamps   ○ 4 groups of 2 stamps

Choose a Nonstandard Unit to Measure Length

Essential Question How can you decide which nonstandard unit to use to measure the length of an object?

Model and Draw

Use  to measure short things.

Use 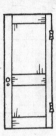 to measure long things.

Share and Show

Use real objects. Circle the unit you would use to measure. Then measure.

	Object	Unit	Measurement
1.			about ____
2.			about ____
3.			about ____
4.			about ____

Math Talk Alex measured a book with ⬭. Then he measured with . Did he use more ⬭ or ➤? Explain.

On Your Own

Use real objects. Choose a unit to measure the length. Circle it. Then measure.

	Object	Unit	Measurement
5.			about _____
6.			about _____
7.			about _____
8.			about _____

PROBLEM SOLVING

9. Fred uses ▣ to measure the stick.
Sue measures the stick and gets the same measurement.
Circle the unit that Sue uses.

TAKE HOME ACTIVITY • Have your child measure something around the house by using small objects such as paper clips and then by using larger objects such as pencils. Discuss why the measurements differ.

Use a Non-Standard Ruler

Essential Question How can you use a non-standard measuring tool to find length?

Model and Draw

About how long is the pencil?

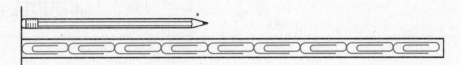

The end of the pencil and the end of the  must line up. Count how many  from one end of the pencil to the other.

about __4__

Share and Show

About how long is the string?

1.

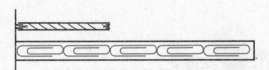

about _____

2.

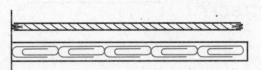

about _____

 Math Talk In Exercise 1, why must the end of the pencil and the end of the  line up?

On Your Own

About how long is the string?

3.

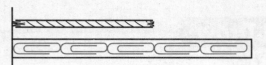

about _____ ⊂⊃

_____

4.

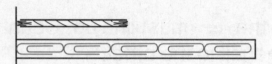

about _____ ⊂⊃

_____

5.

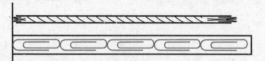

about _____ ⊂⊃

PROBLEM SOLVING REAL WORLD

6. Wendy measures her pencil. She says it is about 2 ⊂⊃ long. Is she correct? Explain.

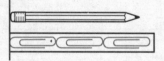

_____

_____

_____

TAKE HOME ACTIVITY • Have your child use 20 paper clips to measure different small objects in your house. Be sure the paper clips touch end to end.

P278 two hundred seventy-eight

Name _____

Compare Lengths

Essential Question How can you compare lengths of objects?

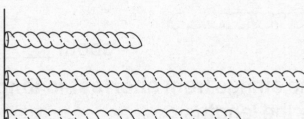
Model and Draw

First, write 1, 2, and 3 to order the strings from **shortest** to **longest**.

Then measure with .

1

3

2

about ___3___ 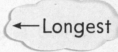 ← Shortest

about ___8___ 🔲 ← Longest

about ___6___ 🔲

Share and Show 🖊️Math Board

Write 1, 2, and 3 to order the strings from **shortest** to **longest**.
Then measure with 🔲. Write the lengths.

1. ___

___

___

about ___ 🔲

about ___ 🔲

about ___ 🔲

 Math Talk How can measuring with cubes tell you the order of the strings?

On Your Own

2. Write 1, 2, and 3 to order the strings from **shortest** to **longest**.
   Then measure with ⬡. Write the lengths.

_____ about _____ ⬡

_____ about _____ ⬡

_____ about _____ ⬡

3. Write 1, 2, and 3 to order the strings from **shortest** to **longest**.
   Then measure with ⬡. Write the lengths.

_____ about _____ ⬡

_____ about _____ ⬡

_____ about _____ ⬡

PROBLEM SOLVING REAL WORLD

4. Kate has these ribbons. Kate gives Hannah the longest one.
   Measure with ⬡ and write the length of Hannah's ribbon.

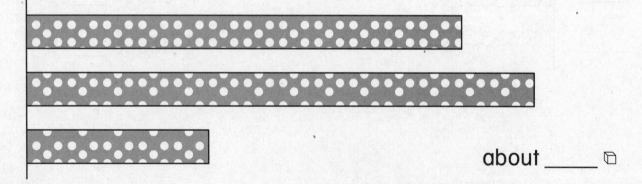

about _____ ⬡

TAKE HOME ACTIVITY • Give your child three strips of paper. Have your child
cut them about 4 paper clips long, about 2 paper clips long, and about 5 paper
clips long. Then have your child order the paper strips from shortest to longest.

Name _____

Time to the Hour and Half Hour

Essential Question How do you tell time to the hour and half hour on an analog clock?

Model and Draw

The hour hand and the minute hand show the time.
Write the time shown on the clock.

4:00

4:30

Share and Show

Read the clock. Write the time.

1.

_____

2.

_____

3.

_____

Math Talk Why does the hour hand point halfway between 5 and 6 at half past 5:00?

On Your Own

Read the clock. Write the time.

4.

_____

5.

_____

6.

_____

7.

_____

8.

_____

9.

_____

PROBLEM SOLVING REAL WORLD

Draw and write to show the time.

10. Liam has soccer practice at half past 10:00.

_____

TAKE HOME ACTIVITY • Say a time, such as half past 1:00 or 7:00. Ask your child where the clock hands will point at that time.

P282 two hundred eighty-two

✓ Checkpoint

Concepts and Skills

Use real objects. Choose a unit to measure the length.
Then measure.

Object	Unit	Measurement
1.		about ____
2.		about ____
3.		about ____

How long is the yarn? Use the star ruler to measure.

4.

____ stars long

5.

____ stars long

Write 1, 2, and 3 to measure the
strings from **shortest** to **longest**.
Then measure with cubes. Write the lengths.

6.

_____   _____ cubes long

_____   _____ cubes long

_____   _____ cubes long

7.

_____   _____ cubes long

_____   _____ cubes long

_____   _____ cubes long

8. Read the clock. Choose the correct time.

○   8:00

○   8:30

○   9:00

○   9:30

Name _____

Use a Picture Graph

Essential Question How do you read a picture graph?

Model and Draw

Our Favorite Hot Dog Toppings					
mustard	♀	♀	♀		
ketchup	♀	♀	♀	♀	♀

Each ♀ stands for 1 child.

_3_ children chose 🫙.

Most children chose ___ketchup___.

_2_ fewer children chose 🫙 than 🍾.

Share and Show

Our Sock Colors						
black	♀	♀				
white	♀	♀	♀	♀	♀	♀
blue	♀	♀	♀			

Each ♀ stands for 1 child.

Use the picture graph to answer the questions.

1. How many children are wearing 🧦? ____

2. What color of socks are most of the children wearing? _____

3. How many more children wear 🧦 than 🧦? ____

Math Talk   How did you find the answer to Exercise 3?

On Your Own

Our Weather						
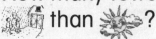 rainy	○	○	○	○		
☀ sunny	○	○				
🌥 cloudy	○	○	○	○	○	○

Each ○ stands for 1 day.

Use the picture graph to answer each question.

4. How many days in all are shown on the graph?

_____ days

5. What was the weather for most days? Circle.

6. How many fewer days were  than 🌥 ?

_____ days

7. How many ☀ and 🌥 days were there?

_____ days

PROBLEM SOLVING

8. Today is sunny. Robin puts one more ☀ on the graph. How many ☀ days are there now?

_____ days

TAKE HOME ACTIVITY • Help your child make a picture graph to show the eye color of 10 friends and family members.

Name _____

Use a Bar Graph

Essential Question How do you read a bar graph?

Fish in the Class Aquarium

Fish		0	1	2	3	4	5	6
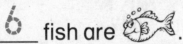 goldfish								
guppy								
angel fish								

Number of Fish

To find how many, read the number below the end of the bar.

___6___ fish are 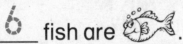.

Share and Show

Use the bar graph to answer the questions.

1. How many fish are in the aquarium?

   _____ fish

2. How many fish in the aquarium are ?

   _____ fish

3. How many fewer fish are  than 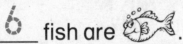?

   _____ fish

4. Are more of the fish 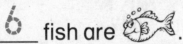 or ?

   _____

 Math Talk   How did you find the answer for Exercise 1?

On Your Own

Use the bar graph to answer the questions.

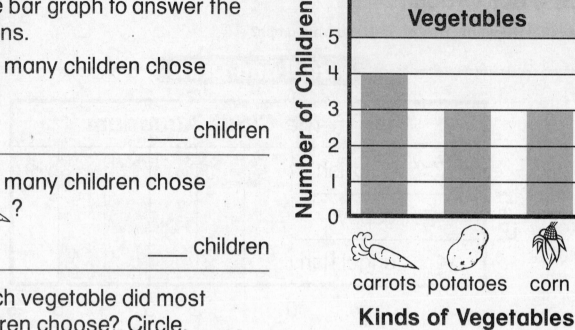

Our Favorite Vegetables

Number of Children

carrots   potatoes   corn

Kinds of Vegetables

5. How many children chose ?

_____ children

_____

6. How many children chose ?

_____ children

_____

7. Which vegetable did most children choose? Circle.

_____

8. Which vegetables were chosen the same number of times? Circle.

_____

PROBLEM SOLVING **REAL WORLD**

Use the bar graph to solve.

9. Brad and Glen both like corn the best. If the boys add this to the graph, how many children will have chosen corn?

_____ children

TAKE HOME ACTIVITY • Ask your child to decide whether they prefer carrots or potatoes. Then have your child color to add their choice to the bar graph on this page.

Take a Survey

Essential Question How can you take a survey?

Model and Draw

You can take a **survey** to get information. Jane took a survey of her friends' favorite wild animals. The tally chart shows the results.

REMEMBER
Each tally mark stands for one friend's choice.

Favorite Wild Animal	
Animal	**Tally**
elephant	
monkey	III
tiger	II

Share and Show

Math Board

1. Take a survey.
Ask 10 classmates which wild animal is their favorite. Use tally marks to show their answers.

_____

Our Favorite Wild Animal	
Animal	**Tally**
elephant	
monkey	
tiger	

2. How many children did not choose tiger?

_____ children

_____

3. Did more children choose elephant or tiger? _____

_____

4. The most children chose
_____ as their favorite.

Math Talk Describe a different survey that you could take. What would the choices be?

On Your Own

5. Take a survey. Ask 10 classmates which color is their favorite. Use tally marks to show their answers.

Our Favorite Color	
Color	**Tally**
red	
blue	
green	

6. Which color was chosen by the fewest classmates? _____

7. Which color did the most classmates choose? _____

8. Did more classmates choose red or green? _____

9. _____ classmates chose a color that was not red.

10. Did fewer children choose blue or green? _____

PROBLEM SOLVING REAL WORLD

11. Jeff wants to ask 10 classmates which snack is their favorite. He makes 1 tally mark for each child's answer. How many more classmates does he need to ask?

Our Favorite Snack	
Snack	**Tally**
pretzels	II
apples	I
popcorn	IIII

_____ more classmates

TAKE HOME ACTIVITY • Have your child survey family members about their favorite sport and make a tally chart to show the results.

Identify Shapes

Essential Question How can attributes help you identify a shape?

Model and Draw

The number of sides and vertices help you identify a shape.

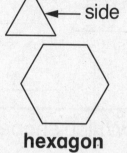
← vertex
← side

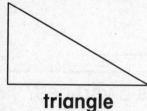
triangle

square

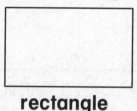
rectangle

trapezoid

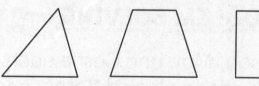
hexagon

3 sides, 3 vertices   4 sides, 4 vertices   6 sides, 6 vertices

Share and Show

Circle to answer the question. Write to name the shape.

1. Which shape has 4 sides?

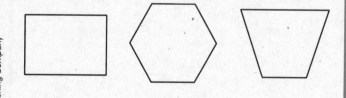

_____

2. Which shape has 3 vertices?

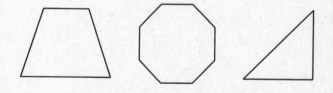

_____

3. Which shape has 6 sides?

_____

4. Which shape has 4 vertices?

_____

Math Talk   How are a square and a rectangle alike?

On Your Own

Circle to answer the question. Write to name the shape.

5. Which shape has 3 sides?

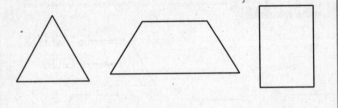

_____

6. Which shape has 4 vertices?

_____

7. Which shape has 4 sides?

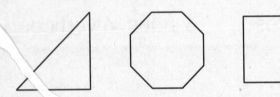

_____

8. Which shape has 6 vertices?

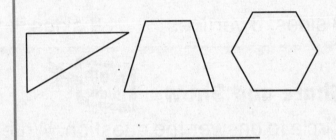

_____

PROBLEM SOLVING REAL WORLD

9. Jason, Mat, and Carrie each draw a shape with 4 sides. The shapes look different and have different names.

Draw 3 shapes the children might have drawn. Write to name each shape.

_____ _____ _____

TAKE HOME ACTIVITY • Have your child look around the house to find something that looks like a rectangle. Then have your child point to the rectangle and count the vertices. Repeat with the sides.

Name _____

Equal Shares

Essential Question How can you name two or four equal shares?

Model and Draw

half	half

__2__ equal shares

__2__ halves

fourth	fourth
fourth	fourth

__4__ equal shares

__4__ fourths

Share and Show

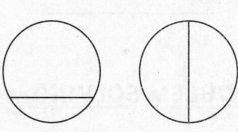

Circle the shape that shows equal shares. Write to name the equal shares.

1.

_____

2.

_____

3.

_____

4.

_____

Math Talk Are all equal shares the same size and shape? Explain.

Getting Ready for Grade 2

On Your Own

Circle the shape that shows equal shares. Write to name
the equal shares.

5.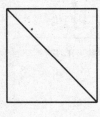

_____

6.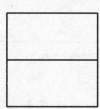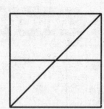

_____

7.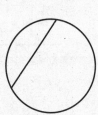

_____

8.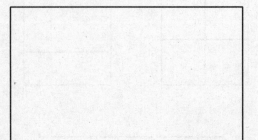

_____

PROBLEM SOLVING REAL WORLD

9. Riley wants to share his cracker with a friend. Draw to show two
different ways Riley can cut the cracker into equal shares.

© Houghton Mifflin Harcourt Publishing Company

 TAKE HOME ACTIVITY • Ask your child to help you cut a piece of toast into
fourths.

Name _____

Concepts and Skills

Use the picture graph to answer Exercises 1 and 2.

Our Favorite Fruit								
🍎 apple	人	人	人	人	人			
🍌 banana	人	人	人	人	人	人	人	人
🍊 orange	人	人	人					

Each 人 stands for 1 child.

1. How many children choose an orange? _____

2. Which fruit was chosen most often? _____

Use the bar graph to answer Exercises 3 and 4.

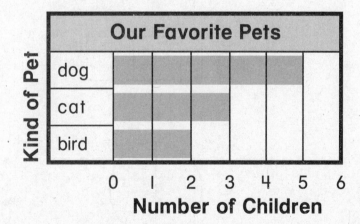

3. Which pet did most children choose? _____

4. How many more children chose a cat than a bird?

_____

5. Take a survey. Ask 8 classmates which sport is their favorite. Use tally marks to show their answers.

Our Favorite Sport	
Sport	Tally
baseball	
football	
soccer	

6. Did more children choose baseball or soccer? _____

Circle to answer the question. Then write the shape name.

7. Which shape has 4 vertices?

  _____

8. Which shape shows fourths?

○   ○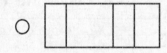

○ 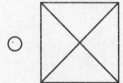   ○

Algebra • Different Names for Numbers

**The blocks show the number in different ways.
Describe the blocks in two ways.**

1. 21

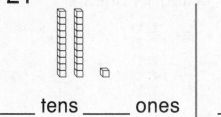

_____ tens _____ ones

_____ + _____

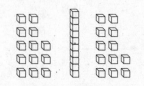

_____ ten _____ ones

_____ + _____

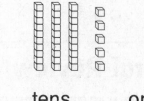

_____ tens _____ ones

_____ + _____

2. 35

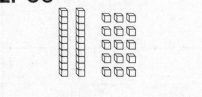

_____ tens _____ ones

_____ + _____

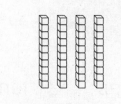

_____ ten _____ ones

_____ + _____

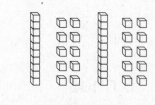

_____ tens _____ ones

_____ + _____

3. 40

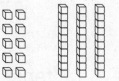

_____ tens _____ ones

_____ + _____

_____ tens _____ ones

_____ + _____

_____ tens _____ ones

_____ + _____

PROBLEM SOLVING **REAL WORLD**

4. Toni has these blocks. Circle
the blocks that she could use
to show 34.

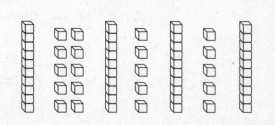

Lesson Check

I. What number is shown with the blocks?

2 tens 9 ones

- ○ 29
- ○ 34
- ○ 30
- ○ 92

2. What number is shown with the blocks?

1 ten 14 ones

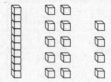

- ○ 14
- ○ 24
- ○ 31
- ○ 36

Spiral Review

3. How many tens and ones make this number? (Lesson 6.4)

16
sixteen

6 ones	1 ten 6 ones	1 ten 7 ones	2 tens
○	○	○	○

4. Which completes the related facts? (Lesson 5.2)

$$3 + 6 = 9 \qquad 9 - 3 = 6$$
$$6 + 3 = 9 \qquad \boxed{}$$

$3 + 3 = 6$	$9 - 6 = 3$	$4 + 5 = 9$	$9 - 4 = 5$
○	○	○	○

Measure with Inch Models

Use color tiles. Measure the length of the object in inches.

1.

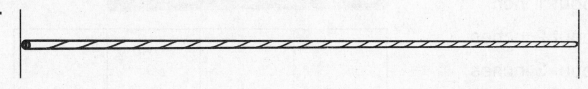

about _____ inches

2.

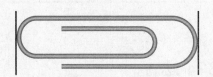

about _____ inches

3.

about _____ inches

4.

about _____ inches

PROBLEM SOLVING

5. Look around your classroom.
   Find an object that is about 4 inches long.
   Draw and label the object.

Lesson Check

1. Jessie used color tiles to measure the rope.
   Which is the best choice for the length of the rope?

   ○ about 1 inch

   ○ about 2 inches

   ○ about 3 inches

   ○ about 4 inches

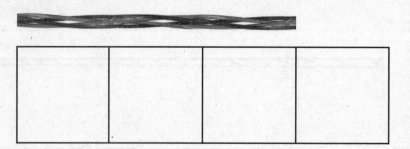

Spiral Review

2. Which doubles fact helps you solve $7 + 6 = 13$? (Lesson 3.5)

   $3 + 3 = 6$ $5 + 5 = 10$ $6 + 6 = 12$ $8 + 8 = 16$
   ○   ○   ○   ○

3. What is the missing number? (Lesson 5.5)

   $7 + \boxed{} = 13$

   ○ 13
   ○ 10
   ○ 6
   ○ 5

4. What is the sum? (Lesson 8.2)

   $30 + 40 = \underline{\hphantom{000}}$

   ○ 70
   ○ 60
   ○ 40
   ○ 10

Make and Use a Ruler

**Measure the length with your ruler.
Count the inches.**

I.

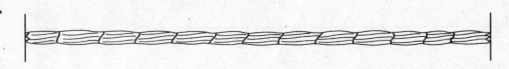

about _____ inches

2.

about _____ inches

3.

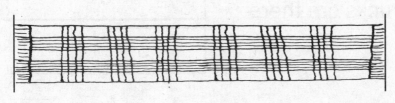

about _____ inches

4.

about _____ inches

PROBLEM SOLVING REAL WORLD

5. Use your ruler. Measure the width
   of this page in inches.

about _____ inches

Lesson Check

1. Use your ruler. What is the best choice for the length of this ribbon?

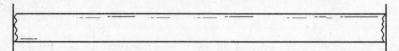

○ about 5 inches

○ about 4 inches

○ about 3 inches

○ about 2 inches

Spiral Review

2. There are 8 red trucks. 5 trucks are blue. How many fewer blue trucks are there than red trucks? (Lesson 2.6)

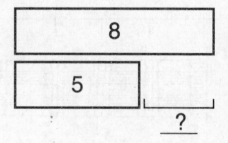

13   8   3   2

○   ○   ○   ○

3. The first group collected 23 cans. The second group collected 34 cans. How many cans did the two groups collect? (Lesson 8.7)

○ 12

○ 43

○ 57

○ 75

4. Which number is greater than 42? (Lesson 7.1)

○ 33

○ 40

○ 42

○ 45

Measure with an Inch Ruler

Measure the length to the nearest inch.

1.

_____ inches

2.

_____ inches

3.

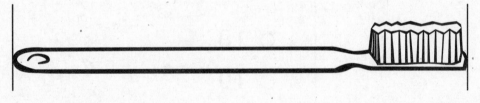

_____ inches

4.

_____ inches

PROBLEM SOLVING REAL WORLD

5. Measure the string. What is its total length?

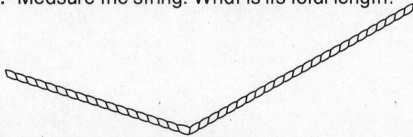

_____ inches

Lesson Check

1. Use an inch ruler. What is the length to the nearest inch?

   ○ 1 inch
   ○ 2 inches
   ○ 3 inches
   ○ 4 inches

2. Use an inch ruler. What is the length to the nearest inch?

   ○ 2 inches
   ○ 3 inches
   ○ 4 inches
   ○ 5 inches

Spiral Review

3. There are 5 big stars and 2 small stars. How many stars are there? (Lesson 3.1)

   ○ 2
   ○ 3
   ○ 5
   ○ 7

4. What is the difference? (Lesson 3.7)

   $$13 - 5 = \underline{\hspace{2cm}}$$

   ○ 18
   ○ 10
   ○ 9
   ○ 8

5. Caleb has 15 pencils
   He gives some away. He has 6 left.
   How many pencils does he give away? (Lesson 4.6)

10	9	7	6
○	○	○	○

Pennies, Nickels, and Dimes

Count by ones, fives, or tens. Write the total value.

1.

_____ ¢   _____ ¢   _____ ¢   _____

2.

_____ ¢   _____ ¢   _____ ¢   _____

3.

_____ ¢   _____ ¢   _____ ¢   _____

4.

_____ ¢   _____ ¢   _____ ¢   _____ ¢   _____

PROBLEM SOLVING  REAL WORLD

5. Eric has some dimes. The total value is 40¢. Draw the dimes Eric has.

Lesson Check

1. What is the total value?

40¢	30¢	20¢	4¢
○	○	○	○

2. What is the total value?

50¢	40¢	25¢	5¢
○	○	○	○

Spiral Review

3. Count back 1, 2, or 3. What is the difference? (Lesson 4.1)

$$8 - 2 = \underline{\hspace{1.5cm}}$$

5	6	7	8
○	○	○	○

4. Which way makes 14? (Lesson 5.8)

7 + 6	3 + 7 + 4	14 + 1	11 − 3
○	○	○	○

Count Collections

Count. Write the total value.

1.

      _____

2.

      _____

3.

      _____

PROBLEM SOLVING REAL WORLD

Draw and label coins to solve.

4. A banana costs 24¢.
   Draw coins you could
   use to buy the banana.

Lesson Check

1. What is the total value?

20¢   22¢   26¢   31¢

○   ○   ○   ○

2. What is the total value?

4¢   31¢   35¢   40¢

○   ○   ○   ○

Spiral Review

3. Which yarn is the longest? (Lesson 9.1)

○

○

○

○

4. What is the sum? (Lesson 3.3)

$$\begin{array}{r} 6 \\ + 6 \\ \hline \end{array}$$

○ 10

○ 11

○ 12

○ 13

Quarters

Count. Write the total value.

1.      _____

2.     _____

3.     _____

PROBLEM SOLVING REAL WORLD

Use coins to show two different ways to make 25¢.
Draw and label the coins.

4.

5.

Lesson Check

1. What is the total value?

- ○ 10¢
- ○ 20¢
- ○ 25¢
- ○ 50¢

2. What is the total value?

4¢	31¢	36¢	46¢
○	○	○	○

Spiral Review

3. What is the sum? (Lesson 8.5)

$$50 + 20 = \underline{\hspace{2cm}}$$

30	52	70	77
○	○	○	○

4. Which is **not** true? (Lesson 7.3)

43 > 40	43 < 40	40 < 43	44 > 43
○	○	○	○

One Dollar

Circle coins to make $1.00.
Cross out the coins you do not use.

1.

2.

3.

PROBLEM SOLVING

4. Draw more coins to show $1.00 in all.

Lesson Check

1. Which group of coins has a value of $1.00?

○   ○   ○

Spiral Review

2. Which addition sentence can you use to check the subtraction? (Lesson 5.4)

$$11 - 3 = \boxed{}$$

$8 + 3 = 11$ $3 + 3 = 6$ $4 + 7 = 11$ $4 + 4 = 8$

   ○   ○   ○   ○

---

3. Count forward. What number is missing? (Lesson 6.1)

$$98, 99, 100, \underline{}, 102$$

96   97   101   103

○   ○   ○   ○